DIETARY FATS, LIPIDS, HORMONES, AND TUMORIGENESIS

New Horizons in Basic Research

ADVANCES IN EXPERIMENTAL MEDICINE AND BIOLOGY

Recent Volumes in this Series

Volume 394
ANTIVIRAL CHEMOTHERAPY 4: New Directions for Clinical Application and Research
Edited by John Mills, Paul A. Volberding, and Lawrence Corey

Volume 395
OXYTOCIN: Cellular and Molecular Approaches in Medicine and Research
Edited by Richard Ivell and John A. Russell

Volume 396
RECENT ADVANCES IN CELLULAR AND MOLECULAR ASPECTS OF ANGIOTENSIN RECEPTORS
Edited by Mohan K. Raizada, M. Ian Phillips, and Colin Sumners

Volume 397
NOVEL STRATEGIES IN THE DESIGN AND PRODUCTION OF VACCINES
Edited by Sara Cohen and Avigdor Shafferman

Volume 398
RECENT ADVANCES IN TRYPTOPHAN RESEARCH: Tryptophan and Serotonin Pathways
Edited by Graziella Allegri Filippini, Carlo V. L. Costa, and Antonella Bertazzo

Volume 399
DIETARY FATS, LIPIDS, HORMONES, AND TUMORIGENESIS: New Horizons in Basic Research
Edited by David Heber and David Kritchevsky

Volume 400A
EICOSANOIDS AND OTHER BIOACTIVE LIPIDS IN CANCER, INFLAMMATION, AND RADIATION INJURY, Part A
Edited by Kenneth V. Honn, Santosh Nigam, and Lawrence J. Marnett

Volume 400B
EICOSANOIDS AND OTHER BIOACTIVE LIPIDS IN CANCER, INFLAMMATION, AND RADIATION INJURY, Part B
Edited by Kenneth V. Honn, Santosh Nigam, and Lawrence J. Marnett

Volume 401
DIETARY PHYTOCHEMICALS IN CANCER PREVENTION AND TREATMENT
Edited under the auspices of the American Institute for Cancer Research

A Continuation Order Plan is available for this series. A continuation order will bring delivery of each new volume immediately upon publication. Volumes are billed only upon actual shipment. For further information please contact the publisher.

DIETARY FATS, LIPIDS, HORMONES, AND TUMORIGENESIS

New Horizons in Basic Research

Edited by

David Heber

University of California, Los Angeles
Los Angeles, California

and

David Kritchevsky

The Wistar Institute
Philadelphia, Pennsylvania

PLENUM PRESS • NEW YORK AND LONDON

Library of Congress Cataloging-in-Publication Data

Dietary fats, lipids, hormones, and tumorigenesis : new horizons in basic research / edited by David Heber and David Kritchevsky.
p. cm. -- (Advances in experimental medicine and biology ; v. 399)
Includes bibliographical references and index.
ISBN 0-306-45317-7
1. Cancer--Nutritional aspects. 2. Lipids in human nutrition. 3. Fatty acids in human nutrition. 4. Hormones--Pathophysiology. I. Heber, David. II. Kritchevsky, David, 1920- . III. Series.
RC268.45.D563 1996
616.99'4071--dc20 96-22307
CIP

Proceedings based on the Nutrition and Cancer Prevention Scientific Symposium, sponsored by the UCLA/NCI Clinical Nutrition Unit, held November 14, 1994, in Los Angeles, California

ISBN 0-306-45317-7

A Division of Plenum Publishing Corporation
233 Spring Street, New York, N. Y. 10013

10 9 8 7 6 5 4 3 2 1

Printed in the United States of America

To Dr. Roslyn B. Alfin-Slater

Professor Emerita
School of Public Health and School of Medicine
University of California, Los Angeles

A scientist, teacher, and prolific writer who dedicated her career to the application of nutritional science to medicine

PREFACE

This book was inspired by a gathering of scientists in Los Angeles in 1994 under the auspices of the UCLA Clinical Nutrition Research Unit which is funded by the National Cancer Institute to promote new research into nutrition and cancer prevention. This unit supports research integrating basic and metabolic/clinical investigations which examine observations from epidemiologic studies and their application to the prevention of common forms of cancer through nutritional intervention.

There is a great deal of information from epidemiologic, experimental and metabolic studies implicating elements of the diet as important in the development and progression of common forms of cancer including breast cancer, colon cancer, prostate cancer, and uterine cancer. When these forms of cancer are examined carefully, it is clear that they share a number of common etiologic factors related to dietary fat, lipids, and hormones.

A human cancer is usually discovered at a point where it has formed a detectable mass. For many forms of cancer, this may require 10 to 15 years from the time when the cancer is first initiated. Nutritional efforts at prevention may delay the progression of cancer to a detectable mass resulting in reduced incidence and may retard the clinical progression and metastatic spread of cancer after its primary treatment.

In humans it is difficult to dissociate the effects of calories from those of fat. In animal model studies, it is possible experimentally to separately consider dietary calories and fat but this is generally not possible in humans under free-living conditions. Free-living humans eating a diet high in fat usually consume more calories than they need. As a result of the intake of additional calories there is an increase in body fat unless physical activity is also markedly increased. However, modern society has developed occupations which involve reduced physical activity if not an entirely sedentary lifestyle. Therefore, obesity has become the foremost nutritional problem in the U.S.today. Obesity has been associated with an increased risk of breast, colon, and prostate cancer in large scale epidemiologic surveys. Obesity is associated with increased circulating levels of reproductive hormones in both men and women. These hormones in turn have been shown to play a role in the promotion of breast and prostate cancers.

Increased intake of fat can also alter the composition of body fat and circulating lipids. These altered patterns may play a role in tumor promotion. Along with an increase in dietary fat intake there is a decrease in the intake of micronutrients including some involved in lipid metabolism. Lipid soluble hormones such as estradiol have long been known to play a role in hormone-dependent cancers. However, there is emerging evidence that lipid-soluble analogs of vitmain D and vitamin A may also play a role in cellular differentiation. The relationship of dietary fat and calories to the development and progression of common forms of cancer is complex. Hormones, lipids, and micronutrients are involved in this process. By its nature nutritional science tends to concentrate on documenting the effects of single nutrients. Integrating this approach with other scientific disciplines has led to a more complex consideration of nutritional mechanisms of cancer prevention.

This book will review evidence demonstrating that dietary fats, lipids and hormones may play an important role in several common forms of cancer. An improved understanding of these factors may lead to new strategies for prevention and treatment based on slowing the rate of disease progression. In some cases, there are opportunities for testing concepts in prospectively randomized intervention trials. It is hoped that this will inspire new avenues of translational research in nutrition and cancer.

D. Heber and D. Kritchevsky

CONTENTS

1

OVERVIEW OF FAT AND CALORIES IN TUMORIGENESIS

David Kritchevsky

The Wistar Institute
3601 Spruce St.
Philadelphia, Pennsylvania 19104-4268

The promoting effect of dietary fat in experimental carcinogenesis was first shown by Watson and Mellanby in 1930 (1). They found that raising the usual 3% fat content of their diet to 12.5-25.0% by addition of butter increased the incidence of skin tumors in coal tar-treated mice by 68%. Several years later Baumann et al. (2) demonstrated that increasing dietary fat (Crisco) from 5 to 25% increased the incidence of skin tumors initiated in mice by ultraviolet irradiation or benzo (a) pyrene (BP) painting. In order to ascertain which fraction of fat was responsible for the increase in tumor formation Lavik and Baumann (3) treated mice topically with methylcholanthrene (MC) and fed 15% hydrogenated vegetable oil, 15% ethyl laurate, 5% glycerol or the unsaponifiable matter obtained from vegetable oil. After 4.5 months tumor yields were 60%, 63%, 47% and 35%, respectively. In the same system heated fat was considerably more tumorigenic (70% incidence at 6 months) then oxidized fat (38%) or irradeated fat (40%). The intact fat (hydrogenated vegetable oil) led to 31% incidence of skin tumors. Their work led Miller et al. (4) to compare the effects of different fats on the carcinogenicity of p-dimethylaminobenzene (DAB) using diets containing 5% fat they compared the effects of corn oil (CO), coconut oil (CNO), hydrogenated coconut oil (HCNO), trilaurin, CO-HCNO 4:1 and CO-HCNO 1:4. The greatest incidence of hepatomas was seen in rats fed CO or CO-HCNO 4:1. A later study (5) compared the effects on DAB-induced hepatomas of 5% corn or olive oil and 20% corn oil, Crisco or lard. Incidence of induced hepatomas at 4 months was: 5% corn oil, 73%; 5% olive oil, 13%; 20% corn oil, 100%; 20% Crisco, 47%; and 20% lard, 60%. Tannenbaum (6) compared the effects of high and low fat diets on BP-induced skin tumors in mice and found that the high fat diet enhanced carcinogenicity and that the effect was most evident during the promotion phase. Further study (7) showed that a high fat diet also led to a higher incidence of spontaneous hepatomas.

Carroll and Khor (8) studied effects of type and quantity of dietary fat on mammary tumors induced in rats by 7,12-dimethylbenz (a) anthracene (DMBA). They showed that increasing dietary corn oil from 0.5 to 5.0% did not affect tumor incidence or multiplicity. Increasing dietary corn oil from 5 to 10% increased tumor incidence by 22% and multiplicity by 43%. Raising the dietary fat level to 20% gave no further increase in either tumor

Dietary Fats, Lipids, Hormones, and Tumorigenesis, edited by Heber and Kritchevsky
Plenum Press, New York, 1996

Table 1. Influence of dietary fat (20%) on incidence of mammary tumors in DMBA-treated rats*

Fat	Incidence (%)	Multiplicity**	Tumorsper 30 rats
Sunflower seed oil	97	4.8±0.31	130
Corn oil	90	4.0±0.62	110
Cottonseed oil	93	4.5±0.75	127
Soybean oil	100	3.4±0.46	103
Olive oil	87	4.5±0.42	117
Lard	93	3.4±0.47	97
Tallow	80	3.0±0.49	72
Butter	87	3.3±0.41	88
Coconut oil	97	2.5±0.38	73

*After Carroll and Khor (8)
**Tumors/tumor-bearing rat.

incidence or multiplicity. Thus some chemical or caloric factor came into play when the diet contained between 5 and 10% corn oil.

The effect of fat saturation was also studied by Carroll and his co-workers (8, 9). They found that saturated fat was less co-carcinogenic than unsaturated fat (Table 1). Similar observations have also been made in the case of 1,2-dimethylhydrazine (DMH)-induced colon tumors (10) and azaserine-induced pancreatic tumors (11). Ip et al. (12) demonstrated a positive essential fatty acid requirement for tumor growth. The requirement reached a maximum at 4.4% linoleate in the diet and plateaued at higher linoleate concentrations. The requirement is for linoleic acid and not for unsaturation which may explain why double bond-rich but linoleate-poor, fish oil inhibits tumor growth (13). Hillyard and Abraham (14) suggest that linoleic acid affects the immune system, possibly via prostaglandin synthesis. In contrast to the foregoing discussion Nauss et al. (14) found no significant effect of fat quantity or saturation on colon cancers induced in rats by N-nitrosomethylurea (NMU) (Table 2). One possible difference may lie in the fact that NMU is a direct-acting carcinogen whereas DMH must be metabolized to provide an active carcinogen.

In the 1930's, McCay and his colleagues (16, 17) reported that male rats fed a severely restricted diet lived 70% longer than ad libitum-fed controls and showed fewer signs of degenerative disease including tumors. Some 30 years earlier Moreschi (18) found that growth of a transplanted sarcoma was significantly reduced in underfed mice (Table 3). Rous (19) showed that underfeeding inhibited growth of both spontaneous and transplanted mouse tumors.

Lavik and Baumann (20) found that the incidence of methylcholanthrene (MCA)-induced skin tumors in mice was zero when the diet was low in both calories and fat. When the diet was high in fat but low in calories tumor incidence was 28% but when it was low in fat and high in calories tumor incidence rose to 54%. A diet high in both calories and fat

Table 2. Lack of effect of fat level on NMU induced colon tumors in rats*

Fat (%)	Incidence (%)	Multiplicity	Mean size (mm ±SD)
Mix (5)**	55	1.64	6.19±3.74
Beef (22): Corn oil (2)	63	1.44	5.56±3.81
Corn oil (24)	58	1.23	7.00±4.71
Shortening (24)	38	1.13	6.29±3.26

*After Nauss et al (15).
**Equal parts of beef, corn oil and shortening

Table 3. Effect of underfeeding on growth of transplanted sarcoma in mice*

Group	Number	Feed intake (g/day)	Weight change (g)	Tumor weight
1	13	1.0	-4.2	1.3±0.2
2	7	1.5	-1.9	3.6±0.5
3	8	2..0	-2.4	5.2±0.5
4	8	ad lib	+1.8	7.6±0.8

*After Moreschi (18).

raised incidence to 66% (Table 4). About a half century later Albanes (21) reviewed 82 studies relating to caloric restriction and tumorigenesis in mice. When the diet was low in both calories and fat (23 studies) tumor incidence was 34 ± 4%; a high fat, low calorie regimen (19 experiments) resulted in an average tumor incidence of 23±5%; diets low in fat but high in calories (18 studies) led to tumor incidence of 52±2%; and when the diet was high in both calories and fat (22 studies) tumor incidence was 54±3%.

In the early 1940's two laboratories in the United States, those of Tannenbaum at the Michael Reese Hospital in Chicago and Baumann at the University of Wisconsin, began to investigate effects of diet on experimental tumorigenesis. Initially, Tannenbaum (22) found that underfeeding reduced the incidence of both spontaneous and chemically-induced tumors in mice. Underfeeding is either designed to maintain the test animals at a predetermined weight or they are fed a proportion of the intake of the control group. Under conditions of underfeeding experiments, if the controls are ingesting the limiting amount of a necessary micronutrient the underfed group may become deficient in this nutrient with deleterious metabolic consequences. In the calorie restriction mode the diets are designed to provide equal levels of vitamins and minerals and restriction is effected at the expense of a macronutrient. After his initial effort, Tannenbaum (23) formulated a diet of known energy content. Energy-restriction studies using his diet showed that incidence of spontaneous mammary or lung tumors or chemically-induced skin tumors was lowered significantly in several strains of mice. Tannenbaum (24) also showed that at the same level of energy intake a diet high in fat led to a higher incidence of mammary tumors than did one low in fat.

Table 5 summarizes some of the early findings relating to caloric restriction or underfeeding on tumorigenesis. A thorough review of the early data was published by White in 1961 (30). It has been suggested that caloric restriction might exert its effect by altering carcinogen metabolism. This may be true in some cases but caloric restriction influences the development of spontaneous tumors as well as tumors caused by ultraviolet irradiation (31) or radioactive calcium (32).

Table 4. Effects of fat and calories on incidence of MCA-induced skin tumors in mice*

Regimen		Tumor incidence
Fat	Calories	%
Low	Low	0
High	Low	28
Low	High	54
High	High	66

*After Lavik and Baumann (20).

Table 5. Influence of caloric restriction on carcinogenesis in mice

Mouse strain	Tumor type (a)	Restriction (%)(b)	Incidence (c)		Ref.
			Control	Restricted	
A	SM	WT	88	16	25
C3H	SM	33	67	0	26
C3H	SM	50	100	13 (V)	27
			100	18 (B)	
DBA	MCL	50	96	35	27
AK	SL	WT	65	10	28
C3H	SM	33	72	0	29

*a) SM = Spontaneous mammary; MCL = Methylcholanthrene-induced leukemia; SL = Spontaneous leukemia.
b) WT = Mice fed enough to maintain starting weight.
c) V = Virgin; B = Breeders

Kritchevsky et al. (33) observed that 40% energy restriction completely inhibited the growth of DMBA induced mammary tumors when the dietary fat was coconut oil (plus enough corn oil for essential fatty acids repletion). The energy -restricted rats were fed twice as much fat as the ad-libitum.-fed controls. Boissonneault et al. (34) administered DMBA to female fats fed on a high-fat diet (300g corn oil/kg), a low-fat diet (50g corn oil.kg) or the high-fat diet restricted by 18.5%. Tumor incidence on the three diets was 73, 43 and 7%, respectively. The daily energy intakes of the three groups were 170, 178 and 146 kJ/d. Klurfeld et al. (35) showed that 40% energy restriction also inhibited DMH-induced tumors in rats. Rats fed on saturated fat exhibited a lower incidence of tumors than rats fed on an unsaturated fat (Table 6).

In a study designed to determine the lowest degree of energy restriction required to inhibit DMBA-induced tumorigenesis. Klurfeld et al. (36) restricted energy by 10, 20, 30 or 40%. All rats, including the controls, ingested 50g fat/kg diet daily. At 10% restriction tumor

Table 6. Influence of 40% caloric restriction and type of fat on colonic and mammary tumors in rats

Regimen	Fat (%)[C]	Tumor incidence (%)	
Colon Tumors[a]			
Ad libitum	Butter oil (3.9)	85	*
Restricted	Butter oil (8.4)	35	
Ad libitum	Corn oil (3.9)	100	
			35
Restricted	Corn oil (8.4)	53	
Mammary Tumors[b]			
Ad libitum	Coconut oil (3.9)	58	
			33
Restricted	Coconut oil (8.4)	0	
Ad libitum	Corn oil (3.9)	80	
			33
Restricted	Corn oil (8.4)	20	

[a] Induced by 1,2-dimethylhydrazine in male F344 rats.
[b] Induced by 7,12-dimethylbenz (a) anthracene in female Sprauge-Dawley rats.
[c] Butter oil and coconut oil diets also contained 1% corn oil. *Unpublished observation

increase in risk with increased caloric intake has not been seen by Stemmerman et al. (57) or Kune et al (58).

EXERCISE

Increasing energy flux may be as effective as reducing energy intake. Rusch and Kline (59) subjected mice bearing a transplanted tumor to enforced exercise (cage rotation) and tumor weight was reduced by about 30%. Vigorous treadmill exercise reduced incidence of DMH-induced colon tumors in rats by about 50%. The incidence of colon tumors in ad libitum-fed, exercised rats was about the same as that observed in sedentary rats subjected to 25% energy restriction (60). Cohen et al. (61) reported that voluntary exercise (activity cage) reduced mammary tumors in rats and voluntary exercise has also been shown to inhibit pancreatic cancer (62). Exercise has also been shown to inhibit growth of Morris hepatoma 7777 in rats (63). On the other hand, Thompson et al. (64, 65) found that treadmill exercise actually enhanced the growth of DMBA-induced mammary tumors in rats. Their exercised rats ate more food and weighed more than the sedentary controls. Vigorous occupational physical activity has been shown to reduce the risk of colon cancer in men (66-68) (Table 10). Risk of cancer may also be reduced by regular exercise (69).

MECHANISMS

The mechanisms by which energy restriction affects carcinogenesis are moot. One probable reason for this is that the effects of energy restriction have not been regarded seriously. Other, more immediate effects on carcinogenesis have been studied in a reflection of the one disease-one cause-one cure paradigm.

Free radicals derived from oxygen have been implicated as possible factors in tumorigenesis and there has been much activity in the cancer field related to anticancer effects of antioxidant vitamins such as A, E and C as well as of carotenoids, Rao et al. (70) showed that energy restriction increased the activity of superoxide dismutase (EC1.15.1.1), catalase (EC1.11.1.6) and glutathione peroxidase (EC 1.11.1.9) in livers of aging rats. Yu (71) reported that the activity of superoxide anion, hydroxyl anion and hydrogen peroxide fell by 27, 28 and 6%, respectively, in the livers of energy-restricted rats. Dietary restriction also increases membrane fluidity (72).

The influences of energy restriction on aspects of DNA metabolism are positive in the direction of cancer control. Energy restriction leads to enhanced DNA repair and reverses

Table 10. Colon cancer risk as related to lifetime work energy expenditure*

Proportion of life spent in sedentary or light work	Group					
	Age 30-65			Age 66-79		
	Cases	Controls	OR[a]	Cases	Controls	OR
None	32	380	1.00	32	209	1.00
1-40%	49	365	1.59	20	85	1.54
41-100%	44	269	1.94[b]	33	123	1.75[c]

*After Vena et al. (67).
a) OR = odds ratio; b) $p<0.01$; c) $p< 0.05$

Table 11. Effects of caloric restriction on plasma insulin levels

Experimental groups	Insulin (mU/ml)	Reference
Ad libitum	122±16	36
30% restricted	42±5	
40% restricted	41±8	
Ad libitum		37
5% fat	143±16	
15% fat	164+15	
20% fat	158±11	
Restricted (25%)		
20% fat	100±12	37
26.7% fat	117±13	
Ad libitum (obese)	1003±193	86
40% restricted (obese)	328±41	

to a degree the age-related loss of specific activity and fidelity of DNA-polymerases (73). Energy restriction increases formation of DNA adducts called I compounds (74) and also aggects DNA adducts of known carcinogens (75).

Djuric and Kritchevsky (76) have reviewed the modulation of oxidative DNA damage by fat and calories. Djuric et al. (77) examined the effects of a low fat diet on oxidative DNA damage in women. DNA damage (expressed as the molar ration of 5-hydroxymethyluracil/10^4 thymine residues in peripheral nucleated blood cells) was reduced significantly ($p<0.0004$) in women on the low fat diet (3.0±0.6 vs. 9.3±1.9). The women on the low fat diet were ingesting 14% fewer calories/day compared to controls. In another study (78) oxidative DNA damage in rats subjected to caloric restriction (43% compared to rats ingesting 5% fat) exhibited 43% less liver DNA damage and 38% less mammary gland DNA damage; both differences were significant. Oncogene expression in rats (79) and mice (80) is reduced by energy restriction as is expression of c-fos and c-ki-ras (81). Effects of dietary restriction on the immune system have been discussed at length by Fernandes and Venkatraman (82).

Insulin deprivation will inhibit tumor growth and cell division (83, 84). When tumor-bearing rats are made diabetic, the tumors stop growing (85). Plasma insulin levels fall significantly in rats subjected to energy restriction (Table 11). Insulin levels fall

Table 12. Mechanisms by which caloric restriction might inhibit tumorigenesis*

1.	Elevated glucocorticoid levels leading to growth inhibition.
2.	Reduction of mitotic activity and cell proliferation.
3.	Prolonged and/or increased immune response.
4.	"Starvation" of preneoplastic cells.
5.	Influence on specific hormones (mammotrophic hormones in the case of mammary tumors.)
6.	Influence on peptide growth factors or receptors.
7.	Modulation and repair of DNA damage.
8.	Influence on oncogene or proto-oncogene expression.

*After Ruggieri (91).

immediately upon institution of energy restriction in rats and remain low throughout the restricted regimen. The levels of insulin-like growth factor I (IGF) also fall at first but rebound to normal within a few weeks; levels of IGF-II are unaffected (87). In mice both insulin receptor and glucocorticoid receptor mRNA increase with age. The latter is not influenced by dietary restriction but the former is increased about 20% in 52% energy restricted mice (88). Hepatic insulin receptors are also increased in energy-restricted rats (89) and binding of insulin to rat liver nuclei is increased by food restriction (90). How these events affect tumor growth is unclear. Ruggeri (91) has summarized the mechanisms by which caloric restriction may inhibit tumorigenesis (Table 12).

The data make it abundantly clear that reduction of energy intake or increased energy output can inhibit spontaneous, induced or transplanted tumors in rodents. The message for man seems clear.

ACKNOWLEDGMENTS

Supported, in part, by a Research Career Award (HL 00734) from the National Institutes of Health.

REFERENCES

1. Watson, A.F. and Mellanby, E. (1930). Tar cancer in mice II. The condition of the skin when modified by external treatment or diet, as a factor in influencing the cancerous reaction. Br. J. Exp. Pathol. 11: 311-322.
2. Baumann, C.A., Jacobi, H.P. and Rusch, H.P (1939). The effect of diet on experimental tumor production. Am. J. Hygiene. 30: 1-6.
3. Lavik, P.S. and Baumann, C.A. (1941). Dietary fat and tumor formation. Cancer Res. 1: 181-187.
4. Miller, J.A., Kline, B.E., Rusch, H.P and Baumann, C.A. (1944). The effect of certain lipids on the carcinogenicity of p-Dimethylaminoazobenzene. Cancer Res. 4: 756-761.
5. Kline, B.E., Miller, J.A., Rusch, H.P. and Baumann, C.A. (1946). Certain effects of dietary fats on the production of liver tumors in rats fed p-Dimethylaminoazobenzene. Cancer Res. 6: 5-7.
6. Tannenbaum, A. (1944). The dependence of the genesis of induced skin tumors on the calorie intake during different stages of carcinogenesis. Cancer Res. 4: 683-687.
7. Silverstone, H. and Tannenbaum, A. (1950). The influence of dietary fat and riboflavin on the formation of spontaneous hepatomas in the mouse. Cancer Res. 11: 200-203.
8. Carroll, K.K., and Khor, H.T. (1971). Effect of level and type of dietary fat on incidence of mammary tumors induced in female Sprague-Dawley rats by 7,12-dimethylbenz (a) anthracene. Lipids 6: 415-420.
9. Gammal, E.B., Carroll, K.K. and Plunkett, E.R. (1967). Effects of dietary fat on mammary carcinogenesis by 7,12-dimethylbenz (a) anthracene in rats. Cancer Res. 27: 1737-1742.
10. Broitman, S.A., Vitale, J.J., Varvrousek-Jakuba, E and Gottlieb, L.S. (1977). Polyunsaturated fat, cholesterol and large bowel tumorigenesis. Cancer 40: 2453-2463
11. Roebuck, B.D., Yager, J.D., Jr, Longenecker, D.S. and Wilpone, S.A. (1981). Promotion by unsaturated fat of azoserine-induced pancreatic carcinogenesis in the rat. Cancer Res. 41: 3961-3966.
12. Ip, C., Carter, CA and Ip, M.M. (1985) Requirement of essential fatty acid for mammary tumorigenesis in the rat. Cancer Res. 45: 1997-2001.
13. Karmali, R.A., Marsh, J. and Fuchs, C. (1984). Effect of omega-3 fatty acids on growth of a rat mammary tumor. J. Natl. Cancer Inst. 73: 457-461.
14. Hillyard, L.A. and Abraham, S. (1979). Effect of dietary polyunsaturated fatty acids on growth of mammary adenocarcinomas in mice and rats. Cancer Res. 39: 4430-4437.
15. Nauss K.M., Locniskar, M. and Newberne, P.M. (1983). Effect of alteration in the quality and quantity of dietary fat on 1.2-dimethylhydrazine-induced colon tumorigenesis in rats. Cancer Res. 43: 4083-4090.
16. Mc Cay, L.M., Crowell, M.F. and Maynard, L.A. (1935). The effect of retarded growth upon the length of life span and upon the ultimate body size. J. Nutr. 10: 63-79.
17. McCay, C.M., Ellis, G.H., Barnes, L.J., Smith, C.A.H. and Sperling, G. (1939). Chemical and pathological changes in aging and after retarded growth. J. Nutr. 18: 15-25.

18. Moreschi, C. (1909). Beziehungen zwischen Ernahrung und Tumorwachstum. Z. fur Immunitatsforsch. 2: 651-675.
19. Rous, P. (1914). The influence of diet on transplanted and spontaneous tumors, J. Exp. Med. 20: 433-451.
20. Lavik, P.S. and Baumann, C.A. (1943). Further studies on the tumor promoting action of fat. Cancer Research 3: 749-756.
21. Albanes, D. (1987). Total calories, body weight and tumor incidence in mice. Cancer Res. 47: 1987-1992.
22. Tannenbaum, A. (1940). The initiation and growth of tumors. Introduction. I. Effects of undernutrition. Am. J. Cancer. 38: 335-350.
23. Tannenbaum, A. (1942). The genesis and growth of tumors. II. Effects of calorie restriction per se. Cancer Res. 2: 460-467.
24. Tannenbaum, A. (1945). The dependence of tumor formation on the composition of the calorie-restricted diet as well as on the degree of restriction. Cancer Res. 5: 616-625.
25. Sivertsen, I. and Hastings, W.H. (1938). A preliminary report on the influence of food and function on the incidence of mammary gland tumor in "A" stock Albino mice. Minnesota Med. 21: 873-875.
26. Visscher, M.B., Ball, Z.B., Barnes, R.H. and Sivertsen, I. (1942). The influence of caloric restriction upon the incidence of spontaneous mammary carcinoma in mice. Surgery 11: 48-55.
27. White, F.R., White, J., Mider, G.B., Kelley, M.G., Heston, W.E. and David, P.W. (1944). Effect of caloric restriction on mammary tumor formation in strain C3H mice and on the response of strain DBA to painting with methyl cholanthrene. J. Natl. Cancer. Inst. 5: 43-48.
28. Saxton, J.A. Jr., Boon, M.C. and Furth, J. (1944). Observations on the inhibition of development of spontaneous leukemia in mice by underfeeding. Cancer Res. 4: 401-408.
29. Huseby, R.T., Ball, Z.B. and Visscher, M.B. (1945). Further observations on the influence of simple caloric restriction on mammary cancer incidence and related phenomena in C3H mice. Cancer Res. 5: 40-46.
30. White, F.R. (1961). The relationship between underfeeding and tumor formation, transplantation and growth in rats and mice. Cancer Res. 21: 281-290.
31. Rusch, H.P., Kline, B.E. and Baumann, C.A. (1945). The influence of caloric restriction and of dietary fat on tumor formation with ultraviolet radiation. Cancer. Res. 5: 431-435.
32. Brown, C.E., Barnes, L.L., Sperling, G. and McCay, C.M. (1960). Radioactive calcium osteosarcoma and squamous carcinomas: Influence of dietary restriction and retarded growth on incidence. Cancer Res. 20: 329-334.
33. Kritchevsky, D., Weber, M.M. and Klurfeld, D.M. (1984). Dietary fat versus caloric content in initiation and promotion of 7,12-dimethylbenz (a) anthracene-induced mammary tumorigenesis in rats. Cancer Res. 44: 3174-3177.
34. Boissonneault, G.A., Elson, C.E. and Pariza, M.W. (1986). Net energy effects of dietary fat on chemically induced mammary carcinogenesis in F344 rats. J. Natl. Cancer Inst. 76: 335-338.
35. Klurfeld, D.M., Weber, M.M. and Kritchevsky, D. (1987). Inhibition of chemically-induced mammary and colon tumor promotion by caloric restriction in rats fed increased dietary fat. Cancer Res. 47: 2759-2762.
36. Klurfeld, D.M., Welch, C.B., Davis, M.J. and Kritchevsky, D. (1989). Determination of degree of energy restriction necessary to reduce DMBA-induced mammary tumorigenesis in rats during the promotion phase. J. Nutr. 119: 286-291.
37. Klurfeld, D.M., Welch, C.B., Lloyd, L.M. and Kritchevsky, D. (1989). Inhibition of DMBA-induced mammary tumorigenesis by caloric restriction in rats fed high fat diets. Int. J. Cancer. 43: 922-925.
38. Welsch, C.W., House, J.W., Herr, B.L., Eliasberg, S.J. and Welsch, M.A. (1990). Enhancement of mammary carcinogenesisby high levels of dietary fat: A phenomenon dependent on *ad libitum* feeding. J. Natl. Cancer Inst. 82: 1615-1620.
39. Tannenbaum, A. (1947). Effects of varying caloric intake upon tumor incidence and tumor growth. Ann. NY. Acad. Sci. 49: 5-17.
40. Weindruch, R. and Walford, R.L. (1982). Dietary restriction in mice beginning at 1 year of age: Effect on life span and spontaneous cancer incidence. Science 215: 1415-1418.
41. Kritchevsky, D., Welch, C.B., and Klurfeld, D.M. (1989). Response of mammary tumors to caloric restriction for different time periods during the promotion phase. Nutr. Cancer 12: 259-269.
42. Ross, M.H. and Bras, G. (1973). Influence of protein under and over nutrition on spontaneous tumor prevalence in the rat. J. Nutr. 103: 944-963.
43. Ross, M.H., Lustbader, E.D. and Bras, G. (1982). Dietary practices of early life and spontaneous tumors of the rat. Nutr. Cancer 3: 150-167.
44. Ross, M.H. and Bras, G. (1971). Lasting influence of early caloric restriction on prevalence of neoplasms in the rat. J. Natl. Cancer Inst. 47: 1095-1113.
45. Hoffman, F.L. (1913). The menace of cancer. Am. J. Obstetrics Dis. Women and Children. 68: 88-91.

46. Berg, J.W., (1975). Can nutrition explain the pattern of international epidemiology of hormone-dependent cancer? Cancer Res. 35: 3345-3350.
47. Lew, E.A. and Garfinkel. L. (1979). Variations in mortality by weight among 750,000 men and women. J. Chronic Dis. 32: 563-576.
48. Garfinkel, L. (1985). Overweight and Cancer. Ann Int Med. 103: 1034-1036.
49. Staszewski, J. (1971). Age at menarche and breast cancer. J. Natl. Cancer Inst. 47: 935-940.
50. DeWaard, F. (1975). Breast cancer incidence and nutritional status with particular reference to body weight and height. Cancer Res. 35: 3351-3356.
51. Apter, D. and Vihko, R. (1983). Early menarche, a risk factor for breast cancer, indicates early onset of ovulatory cycles. J. Clin. Endocrinol. Metab. 57: 82-86.
52. Swanson, C.A., Jones, D.Y., Schatzkin, A., Brinton, L.A. and Ziegler, R.G. (1988). Breast cancer risk assessed by anthropometry in the NHANES I epidemiological follow-up study. Cancer Res. 48: 5363-5367.
53. Kritchevsky, D. (1990). Nutrition and Breast Cancer. Cancer 66: 1321-1325.
54. Jain, M., Cook, G.M., Davis., F.G., Grace, M.G., Howe, G.R. and Miller, A.B. (1980). A case-control study of diet and colorectal cancer. Int. J. Cancer. 26: 757-768.
55. Bristol, J.B., Emmett, P.M., Heaton, K.W. and Williamson, R.C. (1985). Sugar, fat and the risk of colon cancer. Br. Med. J. 291: 1467-1470.
56. Lyon, J.L., Mahoney, A.W., West, D.W., Gardner, J.W., Smith, K.R., Sorenson, A.W. and Stanish, W. (1987). Energy intake: It's relation to colon cancer. J. Natl. Cancer Inst. 78: 853-861.
57. Stemmermann, G.N., Nomura, A.M.Y. and Heilbrun, L.K. (1984). Dietary fat and the risk of colon cancer. Cancer Res. 44:4633-4637.
58. Kune, S., Kune, G.A. and Watson L.F. (1987). Case-control study of dietary etiological factors. The Melbourne Colorectal Cancer Study. Nutr. Cancer. 9: 21-42.
59. Rusch, H.P. and Kline, B.E. (1944). The effect of exercise on the growth of a mouse tumor. Cancer Research 4: 116-118.
60. Klurfeld, D.M., Welch, C.B., Einhorn, E. and Kritchevsky, D. (1988). Inhibition of colon tumor promotion by caloric restriction or exercise in rats. FASEB J. 2: A433.
61. Cohen, L.A., Choi, K. and Wang, C.X. (1988). Influence of dietary fat, caloric restriction and voluntary exercise on N-nitrosomethylurea-induced mammary tumorigenesis in rats. Cancer Res. 48: 4276-4283.
62. Roebuck, B.D., McCaffrey, J. and Baumgartner, K.J. (1990). Protective effects of voluntary exercise during the post initiation phase of pancreatic carcinogenesis in the rat. Cancer Res. 50: 6811-6816.
63. Baracos, V.E., (1989). Exercise inhibits progressive growth of the Morris hepatoma 7777 in male and female rats. Can. J. Physiol. Pharmacol. 67: 864-870.
64. Thompson, H.J., Ronan, A.M., Ritacco, K.A. and Tagliaferro, A.R. (1989). Effect of type and amount of dietary fat on the enhancement of rat mammary tumorigenesis by exercise. Cancer Res. 49: 1904-1908.
65. Thompson, H.J., Ronan, A.M., Ritacco, K.A. and Tagliaferro, A.R. and Meeker, L.D. (1988). Effect of exercise on the induction of mammary carcinogenesis. Cancer Res. 48: 2720-2723.
66. Garabrant, D.H., Peters, J.M., Mack, T.M. and Bernstein, L. (1984). Job activity and colon cancer risk, Am. J. Epidemiol. 199: 1005-1014.
67. Vena, J.E., Graham, D., Zielezny, M., Swanson, M.K., Barnes, R.E. and Nolan, J. (1985). Lifetime occupational exercise and colon cancer. Am. J. Epidemiol. 122: 357-365.
68. Gerhardsson, M., Floderus, B. and Norell, S.E. (1933). Physical activity and colon cancer risk, Int. J. Epidemiol. 17: 743-746.
69. Paffenbarger, R.S. Jr., Hyde, R.T. and Wing, A.L. (1987). Physical activity and incidence of cancer in diverse populations: A preliminary report. Am. J. Clin. Nutr. 45: 312-317.
70. Rao, G., Zia, E., Nadakavukaren, M.J. and Richardson, A. (1990). Effect of dietary restriction on the age-dependent changes in the expression of antioxidant enzymes in rat liver. J. Nutr. 120: 602-609.
71. Yu, B.P., (1991). Free radicals and modulation by dietary restriction. Age and Nutrition. 2: 84-88.
72. Levin, G., Cogan, U. and Mokady, S. (1992). Food restriction and membrane fluidity. Mech. Aging Development. 62: 137-141.
73. Srivastava, V.K., Tilley, R.D., Hart, R.W. and Busbee, D.L. (1991). Effect of dietary restriction on the fidelity of DNA polymerases in aging mice. Exper. Gerontol. 26: 453-466.
74. Randerath, K. (1991). Effects of aging and caloric restriction on I-compounds in liver, kidney and white blood cell DNA of male brown-Norway rats. Mech. Aging Devel. 58: 279-296.
75. Pegram, R.A., Allaben, W.T. and Chou, M.W. (1989). Effect of caloric restriction on aflatoxin B_1-DNA adduct formation and associated factors in Fischer 344 rats: Preliminary findings. Mech. Aging Development. 48: 167-177.

76. Djuric, A. and Kritchevsky, D. (1993). Modulation of oxidative DNA damage levels by dietary fat and calories. Mutation Res. 295: 181-190.
77. Djuric, Z., Heilbrun, L.K., Reading, B.A., Boomer, A., Valeriote, F. and Martino, S. (1991). Effects of a low fat diet on levels of oxidative damage to DNA in human peripheral nucleated blood cells. J. Natl. Cancer Institute 83: 766-769.
78. Djuric, Z., Lu, M.H., Lewis, S., Luongo, D.A., Chen, X.W., Heilbrun, L.K., Reading, B.A., Duffy, P.H. and Hart, R.W. (1992). Oxidative DNA damage in rats fed low-fat, high fat or calorie-restricted diets. Toxicol. Appl. Pharmacol. 115: 156-160.
79. Fernandes, G., Khare, A., Langamere, S., Yu., B., Sandberg, L. and Fredricks, B. (1987). Effect of food restriction and aging on immune cell fatty acids, functions and oncogene expression in SPF Fischer 344 rats. Fed. Proc. 46: 467.
80. Nakamura, K.D., Duffy, P.H., Lu, M.S., Turturro, A. and Hart, R.W. (1989). The effect of dietary restriction on myc protooncogene expression in mice: A preliminary study. Mech. Aging Develop. 48: 199-205.
81. Himeno, Y., Engelman, R.W. and Good, R.W. (1992). Influence of calorie restriction on oncogene expression and DNA synthesis during liver regeneration. Proc. Natl. Acad. Sci. (USA) 89: 5497-5501.
82. Fernandes, G. and Venketraman, J.T. (1993). Dietary restriction. Effects on Immunological Function and Aging. In: Human Nutrition: A Comprehensive Treatise. Vol. 8 Nutrition and Immunology (Ed. D.M. Klurfeld) New York, Plenum Press pp 91-120.
83. Cohen, N.D. and Hilf, R. (1974). Influence of insulin on growth and metabolism of 7, 12-dimethylbenz (a)-anthracene-induced mammary tumors. Cancer Res. 34: 3245-3252.
84. Taub, R., Roy, A., Dieter, R. and Koontz, J. (1987). Insulin as a growth factor in rat hepatoma cells. J. Biol. Chem. 262: 10893-10897.
85. Heuson, J.C. and Legros, N. (1972). Influence of insulin deprivation on growth of the 7, 12-dimethylbenz (a)-anthracene-induced mammary carcinoma in rats subjected to allozan diabetes and food restriction. Cancer Res. 32: 226-232.
86. Kritchevsky, D. (1990). Influence of caloric restriction and exercise on tumorigenesis in rats. Proc. Soc. Exp. Biol. Med. 193: 35-38.
87. Ruggeri, B.A., Klurfeld, D.M., Kritchevsky, D. and Furlanetto, R.W. (1989). Caloric restriction and 7, 12-dimethylbenz (a)anthracene-induced mammary tumor growth in rats: Alterations in circulating insulin, insulin -like growth factors I and II and epidermal growth factor. Cancer Res. 49: 4130-4134.
88. Spindler, S.R., Grizzle, J.M., Walford, R.L. and Mote, P.L. (1991). Aging and restriction of dietary calories increases insulin receptor mRNA and aging increases glucocorticoid receptor mRna in the liver of female C3B10RF mice. J. Gerontol. 46: B233-237.
89. Balage, M., Manin, M., Arnal, M. and Grizard, J. (1988). Differential regulation of muscle and liver insulin receptors by energy restriction in growing rats. Reproductive Nutr. Devel. 18: 819-820.
90. Venkatraman, J. and Fernandes, G. (1992). Modulation of age-related alterations in membrane composition and receptor-associated immune functions by food restriction in Fischer 344 rats. Mech. Aging Devel. 63:27-44.
91. Ruggeri, B.A., (1991). The effect of caloric restriction on neoplasia and age-related degenerative processes. In: Human Nutrition: A Comprehensive Treatise. Vol. 7. Cancer and Nutrition (Eds. R. B. Alfin –Slater and D. Kritchevsky), New York Plenum Press, pp. 187-210.

2

INTERRELATIONSHIPS OF HIGH FAT DIETS, OBESITY, HORMONES, AND CANCER

David Heber

Division of Clinical Nutrition
UCLA School of Medicine
1000 Veteran Avenue
Los Angeles, California 90095

INTRODUCTION

Obesity is defined as excess body fat and results from an imbalance of energy intake and expenditure (1). This simple definition and the application of thermodynamic principles of energy balance do not adequately express the fact that much remains unknown about this complex disorder (2). In common with other chronic diseases including cancer, it is rare for obesity to be the expression of a pure genetic or medical disorder. In the majority of cases, obesity results from a combination of physiological, psychological and environmental factors (3). A genetic predisposition to obesity mediated by a large number of genes is likely to interact with a permissive high fat diet in the presence of reduced physical activity to result in obesity(4). Obesity is the most prevalent disorder in the United States affecting up to one-third of the population. The endocrine system which is critically involved in many common forms of cancer has also been shown to translate environmental influences into excess adiposity (5). Many hormones involved in obesity also play a role in the initiation and promotion of cancer both at a cellular, paracrine, and systemic level (6). For example, reproductive hormones are critically involved in both obesity and cancer. In patients with primary anorexia nervosa and malnutrition, reproductive hormone secretion is reduced and can result in infertility. Conversely, with obesity excess secretion of reproductive hormones (e.g. estrogens) has been observed and implicated in the etiology of breast cancer (7). This chapter will examine the potential role of hormones in the etiology of obesity in the hopes of stimulating additional research on the role of obesity in modulating the risk of developing several common forms of cancer.

EPIDEMIOLOGY OF OBESITY

It has been proposed that the fat cells of individuals living in societies characterized by a low intake of fat, high intake of fiber and micronutrients and high levels of physical

Dietary Fats, Lipids, Hormones, and Tumorigenesis, edited by Heber and Kritchevsky
Plenum Press, New York, 1996

by common infectious diseases (8). Since immune function is seriously impaired with malnutrition (9), the ability of individuals to avoid malnutrition in conditions of food scarcity would be expected to associated with an increased survival of the obese genetic trait. This is particularly exemplified in the case of the Pima Indians living in Arizona(10). This group of Native Americans is genetically identical to a group living in Mexico under conditions which have remained unchanged for hundreds of years. In Mexico there is little obesity or chronic disease, while among the Pimas in Arizona there is a high incidence of diabetes, hypertension, and obesity. The Pima Indians living in Arizona have eaten a high fat Westernized diet since a famine in the early part of this century resulted in the loss of their traditional diet and lifestyle.

Obesity has been shown to increase the risk for development of several common cancers including those affecting the breast, prostate, colon, and uterus.

It is possible that two chronic conditions (i.e. obesity and cancer) are simply associated by chance, but there are a number of potential interactions between these two conditions which could lead to a causal interrelationship. In obesity, the circulating level of certain sex steroid hormones including estradiol are increased. These increased levels might promote the development of hormone-dependent cancers of the breast. In addition, these tumors have receptors for peptide growth factors including insulin-like growth factor I. It is also known that the levels of this peptide growth factor are increased in obesity. Therefore, obesity may result in increased circulating levels of growth factors which promote cancer development and progression. There is also the possibility that adipose tissues have a paracrine relationship to certain epithelial tumors such as breast cancer.

HORMONES AND OBESITY

In the obese patient, changes in hormone secretion and action result both from the effects of ongoing positive caloric balance and excess adiposity. In the course of weight reduction, the levels of many hormones are decreased in a biphasic manner with an initial decrease secondary to caloric restriction over the first one to two weeks of dieting followed by a further decline as weight reduction progresses over several months. A number of hormones can promote obesity, and some hormones have been utilized in the treatment of obesity. It is clear that changes in endocrine function can affect both the assessment and treatment of obesity.

It is a common misconception that obesity is most often a glandular disorder resulting from a deficiency of a particular hormone which promotes normal nutrient metabolism. Obese patients will frequently request a complete endocrinological examination to exclude this possibility.

HYPERINSULINEMIA

There are several conditions where hyperinsulinemia is seen and has been implicated in the etiology of associated obesity. The most important association of obesity in this regard is with noninsulin-dependent or type II diabetes mellitus. Most, but not all patients with this type of diabetes are obese (11) and these patients comprise the majority of the approximately 15 million diabetics in this country (12). Obese non-diabetic subjects also have hyperinsulinemia(13). This has been shown to be secondary to both increased insulin secretion from the pancreas and decreased hepatic clearance of insulin (14) especially in individuals with abdominal or central obesity. In addition, obese patients are resistant to the effects of insulin on glucose metabolism with lesser effects on lipid and protein metabolism (15). Therefore, higher insulin levels result in part from the abnormal control of glucose metabolism. These higher insulin levels promote lipogenesis via increased activity of lipoprotein lipase, the enzyme which catalyzes the hydrolysis of circulating triglycerides into fatty acids and glycerol for uptake by the adipocyte to synthesize stored triglyceride intracellularly (16). In the obese population, hyperglycemia is uncommon, but the exact size of the latent diabetic population or the population which might benefit from weight reduction by an amelioration or delay in the development of obesity remains to be determined.

Iatrogenic hyperinsulinemia can result from the administration of excessive insulin in an attempt to regulate blood glucose levels in diabetic patients who do not adhere to dietary therapy. In these patients, increased insulin administration is associated with increased hunger and weight gain. The same phenomenon can be observed with excessive use of oral hypoglycemic agents. Hyperinsulinemia at different levels of blood glucose achieved using the euglycemic insulin clamp technique is associated with increased hunger, heightened perceived pleasantness of sweet taste and increased food intake (17).

For many common forms of cancer insulin and insulin-like growth factors can promote tumor growth. The evidence for breast cancer is further discussed in the chapter on stromal cell-breast cancer interactions.

POLYCYSTIC OVARIAN SYNDROME (STEIN–LEVENTHAL SYNDROME)

This heterogeneous syndrome is characterized by infertility, dysmenorrhea ,hirsutism, and insulin resistance (18). Insulin resistance in these patients has been associated with both increased ovarian androgen and pituitary gonadotropin secretion in this syndrome (18). While some epidemiologists have suggested that this syndrome may be a risk factor for ovarian cancer, there is little evidence as yet favoring this association.

REPRODUCTIVE HORMONES AND WEIGHT GAIN

Approximately 80 percent of all patients in weight reduction programs are women. The effects of reproductive hormonal status on fat accumulation and mobilization are critical issues in the understanding of the pathogenesis of obesity and common forms of cancer in women.

Following the onset of puberty, women have increased adiposity compared to men. Obese women have higher levels of circulating estrogens than lean women due to conversion of adrenal androgens to estrogens by an aromatase enzyme found in the adipose stromal cells (19). Premenopausal women develop gluteofemoral fat as well as abdominal fat, while men develop primarily abdominal fat (20). The gluteofemoral fat is a useful store of energy for lactation and increases during each pregnancy. However, in our society women rarely feed their infants at the breast to the extent found in more primitive societies so that the gluteofemoral fat is not lost effectively (23). In fact, food intake can increase after pregnancy, and post-pregnancy weight gain is an important factor in the epidemiology of obesity in American women (24). Premenopausal women have two significant sources of estrogen, the ovaries and their fat tissue. Following the menopause, the gluteofemoral fat depots decrease in size in many women while adiposity increases in the abdominal and breast fat depots. The use of pharmacological amounts of estrogens by males who impersonate females leads to gynoid fat distribution, while the use of anabolic androgens by female weightlifters leads to male fat patterns. While these data suggest that the gluteofemoral fat depots are estrogen-dependent, estrogen receptors have not been found in cells isolated from gluteofemoral fat (25).

The activity of lipoprotein lipase (LPL) in gluteofemoral fat is affected by reproductive endocrine status. Gluteofemoral fat in premenopausal women has higher levels of LPL activity than abdominal fat when normalized per surface area (26). During early pregnancy (up to 10 weeks) this difference in LPL activity is more pronounced. In late pregnancy LPL activity begins to decrease in the femoral adipocytes. During lactation, LPL activity declines further in the femoral adipocytes until there is little difference from the levels found in the abdominal adipocytes. These changes parallel the storage and subsequent mobilization of fat from the gluteofemoral fat stores. In early pregnancy and in non-pregnant premenopausal women, the abdominal adipocytes with a lower LPL activity are found to release fatty acids more readily in response to stimulation by norepinephrine than gluteofemoral adipocytes(27). In late pregnancy and during lactation, triglycerides are as easy to mobilize from gluteofemoral as from abdominal fat providing energy and fat for lactation. Women report craving high fat and sweet foods during the last 10 days of the menstrual cycle when circulating progesterone levels are significantly elevated compared to the follicular phase, and this has been documented in nutritional studies of premenopausal women (28). A progestational steroid drug, megestrol acetate, has been shown to increase appetite in malnourished cancer and AIDS patients with resultant significant weight gain (29).

It is well-established that age at menarche, age at menopause, parity and age at first full-term pregnancy are important determinants of breast cancer risk (30). Asian women living a traditional lifestyle have lower estrogen levels both before and after menopause (31), and these lower levels of estrogen have been considered important markers of the differences in breast cancer incidence observed in international epidemiological studies (32). Increased daily alcohol intake has been associated with increased risk of breast cancer and increased levels endogenous estrogens in premenopausal women (33). A number of aspects of dietary intake, physical activity and body composition affect endogenous estrogen levels. In our own studies of over 300 women, a decrease in fat intake was associated with a significant decrease in body fat measured by bioelectrical impedance (34). We have also shown that a high fiber low fat diet will reduce serum estrone and estradiol in both the follicular and luteal phases of the cycle (35). Dietary fat and fiber intake have also been reported to affect the circulating concentrations of estrogens and androgens in women eating their usual diets under controlled conditions(36).

BODY FAT, DIETARY FAT AND ESTROGENS IN BREAST CANCER PREVENTION

Increased estrogen production in women with excess fat compared to lean women (37,38) is due to increased peripheral conversion of androstenedione to estrone by fat tissue (39,40). Obesity has been associated repeatedly with more advanced breast cancer at the time of diagnosis, higher rates of recurrence, and shorter survival times even after controlling for tumor size and stage of disease at diagnosis (41-46). This has been explained in the past as resulting from delayed diagnosis in obese women rather than enhanced tumor promotion resulting from increased estrogen production. The effects of obesity on the invasiveness of breast cancers at the time of diagnosis are different among patients with estrogen receptor-positive tumors compared to those with estrogen receptor-negative tumors (47). If the effects of obesity on breast tumor nodal invasion and, therefore, prognosis were due simply to delayed diagnosis then similar associations should have been seen for both the receptor-positive and negative patients. The fact that such an effect of obesity was observed only in the receptor positive patients strongly suggests that a hormonal mechanism accounts for the effects of obesity on breast tumor promotion.

While obesity on a weight for height basis has not been associated with an increased risk of premenopausal breast cancer, adult weight gain increases the risk of postmenopausal breast cancer and is a predictor of breast cancer risk independent of body weight (48-58). The physiological effects of weight gain are likely to be evident in premenopausal as well as postmenopausal women regardless of whether they are classified as obese on weight for height criteria.

Our own studies demonstrated a decrease in estradiol and estrone levels with no change in estrone sulfate in 12 premenopausal women following 8 weeks on a 10% fat, high fiber diet by comparison to levels measured following one month of a 30% fat diet. The primary focus of this work was to demonstrate that estrogen levels could be reduced without changing menstrual function in normal premenopausal women. Subjects first participated in a screening month during which they were asked to monitor their basal body temperature in order to determine the approximate length of the menstrual cycle and to confirm normal ovulation. Basal body temperatures were monitored throughout the four months of the study. During the second month subjects received one meal at the UCLA General Clinical Research Center (GCRC) and followed a diet providing 30% fat calories and 15-20 gm fiber/day. During months 3 and 4 subjects received all their main meals at the GCRC and were instructed to record any other food they ate carefully. A fourteen day menu cycle was devised and reviewed with each subject. The very low fat, high fiber (VLF/HFi) diet provided 10% fat calories and 25 to 35 gm fiber/day. Blood samples were drawn every other day to monitor the levels of reproductive hormones. During controlled feeding of both diets subjects were provided with more than adequate food and were asked to return unused food which was weighed to assess the amount of food eaten. Four day food records were kept by all subjects and analysed using the University of Minnesota Nutrient Data Base System established in the UCLA CNRU.

Subjects were weighed weekly in the GCRC. There was a statistically significant overall loss of body fat and body weight in the 12 women studied as shown in Table 1.

As shown in tables 2 and 3, there were significant reductions in plasma estradiol and estrone with the high fiber, low fat diet without any significant effect observed on menstrual

Table 1. Nutrient intake based on four day food records

	Diet Protocol**		
	30% AHA	VLF/HFi	VLF/HFi
Variable (mean± SD)	Month 1	Month 2	Month 3
Energy (kcal)	1524 ± 282	1462 ± 159	1520 ± 201
Protein (gm)	73.4 ± 17.4	66.6 ± 9.4	68.8 ± 14.4
Fat(gm)	49.1 ± 16.4	17.2 ± 4.8*	20.2 ± 4.6*
Carbohydrate(gm)	205.09 ± 27.2	272.5 ± 32.6*	280.0 ± 30.4*
Protein(%)	19.7 ± 2.85	18.38 ± 2.4	17.9 ± 1.6
Fat(%)	28.5 ± 4.8	10.5 ± 2.7*	11.9 ± 1.7*
Carbohydrate(%)	54.0 ± 3.9	74.5 ± 3.4*	74.0 ± 2.1*
Cholesterol(mg)	164.4± 48.6	59.5 ± 25.2*	73.8 ± 32.9*
P:S ratio	0.84	0.93	0.93
Fiber (gm)	18.9 ± 5.5	28.9 ± 3.4*	27.9 ± 4.9*

* Significantly different from 30% AHA phase, $p<0.01$

** AHA = American Heart Association 30% fat diet

VLF/HFi = very low fat (10% en), high fiber (25-35 gm/day) diet

Table 2. Effects of VLF/HFi diet on body weight

	Diet Protocol**	+	
	30% AHA Diet	VLF/HFi	
	Month 1	Month 2	Month 3
Weight(kg)	61.7 ± 8.6	60.7 ± 8.7*	59.7 ± 9.0*
Height(m)	1.66 ± .09	1.66 ± .09	1.66 ± .09
BMI(kg/m^2)	22.3 ± 1.7	21.9 ± 1.9*	21.5 ± 5.5*
% Body fat	23.5 ± 4.9	24.0 ± 6.6*	21.6 ± 5.5*

* Significantly different from 30% AHA phase, $p<0.01$
**AHA = American Heart Association 30% fat diet
VLF/HFi = very low fat (10% en), high fiber (25-35 gm/day) diet

cycle length. In the above studies, the effects on menstrual cycle length were observed in only 12 women over two menstrual cycles.

Reichmann et. al. (59) demonstrated that low fat (20%) lengthened follicular phase length in 20 of 27 women compared to a higher fat (40%) diet. In that study, 4 women had a shortening of follicular phase length and 3 had no change. In addition to our own work which is shown above, Rose et.al.(60) instructed 16 premenopausal females with benign breast disease to reduce fat consumption from 69 g to 32 g of fat per day and found a significant (ca.30%) reduction in estradiol levels after 3 months on the diet. Rose et.al.(61) also reported a decrease in luteal phase estradiol and estrone levels with an increase in fiber intake from 15 to 30 g per day. Woods et. al. demonstrated a decrease in estrone sulfate concentrations when premenopausal women were fed a 20% fat high fiber diet (62).

Bernstein et al. in a case-control study of healthy premenopausal women and breast cancer patients in Shanghai and Los Angeles found that Los Angeles controls had a 20.6% greater estradiol concentration than Shanghai controls and adjustment for body weight accounted for only 25.7% of this difference(63). They concluded that the higher estrogen levels could be an important part of the two to three fold difference in breast cancer incidence rates of women under age 45 in Shanghai and Los Angeles. Of interest, the dietary intervention we have tested results in decreases in estradiol levels of similar magnitude to the observed differences in this epidemiologic study. Additionally, there were no significant changes in progesterone levels in our study or in the observations made by Bernstein et al. when the subset of ovulating premenopausal women were considered in her study.

Significant changes in plasma lipids were observed in these subjects after one month on the diet as shown in Table 4. It should be noted that these decreases from normal levels were quite marked by comparion to our earlier studies of the effects of a 20% fat diet in 300

Table 3. Effects of VLF(10%)/HFi(35-45 g/day) diet on menstrual cycle length

	Diet Protocol**		
	Control (30% AHA)	Month 1 VLF/HFi	Month 2 VLF/HFi
Menstrual Cycle Length (days)	28.2 ± 3.4	28.2 ± 4.2	25.8 ± 5.2
Follicular Phase (days)	15.8 ± 4.2	16.2 ± 3.7	13.4 ± 4.1
Luteal Phase (days)	12.3 ± 3.0	12.0 ± 1.7	12.4 ± 3.0

**AHA = American Heart Association 30% fat diet
VLF/HFi = very low fat (10% en), high fiber (25-35 gm/day) diet

Table 4. Effects of VLF(10%)/HFi(35-45 g/day) diet on plasma estradiol and estrone

	Diet Protocol**		
	Control (30% AHA)	Month 1 VLF/HFi	Month 2 VLF/HFi
Estradiol (follicular)	73.4 ± 16.7	60.2 ± 19.1	54.8 ± 29.9 * [-26%]
Estradiol (luteal)	193 ± 86	211 ± 84	151 ± 70* [-22%]
Estrone (follicular)	73.4 ± 27	70.1 ± 40.5	59.4 ± 23.7* [-20%]
Estrone (luteal)	86.7 ± 35	72.6 ± 27	71.4 ± 32.9* [-18%]

AHA = American Heart Association 30% fat diet
VLF/HFi = very low fat (10% en), high fiber (25-35 gm/day) diet

postmenopausal women where cholesterol was weakly related to weight loss ($p<0.01$, $r=0.24$) and to self-reported dietary adherence ($p<0.05$).

In a study our group conducted with 311 postmenopausal women undergoing a dietary intervention patterned after a national multicenter protocol (the Women's Health Trial), total reported calorie intake was also reduced from 1842 kcal/day to 1413 kcal/day at 8 weeks of treatment with maintenance at 1420 kcal/day at 1 year. In common with similar studies, there was a small average weight loss on this program of approximately 2 kg. in 238 women after one year of participation in the program. In these women, body mass index decreased from 26.15 to 25.4 kg/sq.m., and percent body fat measured by bioelectrical impedance decreased from 30.9 percent at baseline to 27.6 percent at one year. Therefore, low fat, high fiber diets such as the one utilized in this study should cause significant changes in body composition especially when combined with exercise.

Breast ductal fluid(BDF) total fatty acid concentrations were changed by the dietary intervention. The levels of many fatty acids changed, but the changes in linoleic acid which decreased in the serum from 2318 ± 270 to 1910 ± 247 umol/L (-18%,mean ± SD,$p<0.05$) were accompanied by a decrease in BDF linoleate from 36265 ± 16524 to 28055 ± 15304 umol/L (-23%, mean ± SD, $p<0.05$). These data demonstrated that the dietary intervention affected the breast ductal cell microenvironment. Total cholesterol, triglycerides, total estrogens, and progesterone did not change. Unfortunately, the BDF sample was only obtained before and one month after initiating the diet prior to changes in serum levels of hormones. Therefore, while no changes were noted in total estrogens in the BDF at one month, it is possible that these levels may change with more prolonged controlled feeding.

Recently, 28 women between the ages of 25 and 45 seen in the UCLA Breast Center's High Risk Clinic underwent measurement of weight, height, and percent body fat by bioelectrical impedance. While there are no normalized tables for lean body mass, the percent body fat in women should ideally be between 22 and 28%. In this clinic population only 2 of 28 women were in the normal range. These data shown in table 6 strongly suggest that

Table 5. Changes in plasma lipids and fatty acids

	Diet Protocol*	
	Baseline	VLF/HFi* 1 month
Plasma Total Cholesterol (mg/dL)	173 ± 24.5	152 ± 25*
Plasma LDL Cholesterol (mg/dL)	104 ± 16.5	83 ± 17*
Plasma Triglycerides (mg/dL)	89 ± 25	103 ± 25
Plasma Total Linoleate (umol/L)	2318 + 270	1907 + 327*

* VLF/HFi = very low fat (10% en), high fiber (25-35 gm/day) diet

Table 6. Body mass and percent body fat in women at increased risk of breast cancer

	Age (yr)	Wt. (lbs)	Ht. (in)	BMI (wt/ht^2)	Body Fat (%)
Mean ± SD (n=28)	36.8±6.4	37.8±1.9	165.3±2.7	22.9±3.1	34.6±4.8

these women had decreased lean body mass, since their body weight for height as assessed by Body Mass Index was in the normal range.

HYPOGONADISM

Hypogonadism is associated with increased body fat in both men and women(64). The exact mechanisms underlying the accumulation of eunuchoid fat is not known, but the low level of testosterone relative to estrogen has been raised as a possible mechanism. In men, the low levels of circulating estrogen result primarily from peripheral aromatization of adrenal androgens by fat tissue. Finally, growth hormone deficiency is associated with an accumulation of body fat (65). There is approximately a 2.5 kg average weight gain observed at the time of the menopause in women (66). It has not been determined whether the observed weight gain is due to a metabolic effect of the loss of estrogen and progesterone, decreased physical activity or simply to increased food intake at the time of the menopause.

ENDOCRINE CHANGES SECONDARY TO OBESITY

Endocrine function is markedly affected by obesity. In some cases, the differences observed in obese compared to lean subjects have clear implications for the etiology of obesity as discussed above for insulin and estrogen secretion. In other cases, the differences in function do not have clear consequences for the physiological changes observed in obese subjects. However, abnormalities secondary to obesity are important in understanding the potential effects of obesity on cancer development, promotion, and progression.

INSULIN RESISTANCE

As discussed above, obese patients commonly have hyperinsulinemia. However, despite the high levels of insulin observed, such patients do not become hypoglycemic suggesting that they are resistant to the hypoglycemic effects of insulin. Exogenous insulin administration leads to a subnormal drop in blood glucose levels in obese patients (67). This insulin resistance is most pronounced at physiological and submaximal doses of insulin. The response to maximal doses of insulin infused under euglycemic conditions appears to be normal. In addition, the resistance to insulin effects on amino acid metabolism and lipogenesis appear to be insignificant. The latter point is important since insulin stabilizes LPL post-translationally promoting lipogenesis in the adipocyte and promoting obesity in the presence of insulin resistance. Once glucose homeostasis fails, then subjects with insulin resistance develop non-insulin dependent diabetes mellitus as already discussed in detail. The interrelationships of hyperinsulinemia, insulin resistance and the etiology of obesity are closely intertwined and it is not possible to state which is primary and which is secondary.

DECREASED GROWTH HORMONE SECRETION

Growth hormone secretion is blunted in obese patients both when determined in the fasting state or over a 24 hour period (68). Increases in growth hormone secretion in response to a number of stimuli including administration of GHRH (growth hormone releasing hormone)(69), sleep (70), fasting (71), protein-containing meals (72), or exercise (73). Growth hormone secretion can be increased in obese patients by administration of thyroid hormone(74), by fasting(75), or by pulsatile administration of GHRH(76).

It is likely that the defects in growth hormone secretion are secondary to obesity, but are not involved in the etiology of obesity. First, lean subjects who are overfed develop impaired growth hormone responses to a number of stimuli including sleep, exercise, hypoglycemia or GHRH administration (77). Second, when obese subjects lose weight normal growth hormone responsiveness is usually restored (78). Finally, the degree of impairment of growth hormone responsiveness appears to be related to degree of obesity (79).

In many tissues growth hormone promotes growth indirectly through stimulation of somatomedin or IGF-1 (insulin-like growth factor 1) secretion. In obese subjects, IGF-1 levels are normal or elevated in the face of reduced growth hormone levels explaining why growth is normally maintained in obese children despite low levels of growth hormone (80). The mechansim by which IGF-1 levels are maintained at normal or elevated levels has not been established. However, height at the time of puberty has been implicated as a risk factor for breast cancer. It is known that when food intake is increased in a population such as that in Japan over the past twenty years that there is an increase in average height.

It is not known whether the abnormalities in growth hormone secretion or responsiveness to physiological and pharmacological stimuli have important consequences for the obese patient or are merely interesting epiphenomena. However, it may be important to consider these changes in studies of nutrition and cancer.

BLUNTED PROLACTIN RESPONSIVENESS

The serum levels and 24 hour integrated serum prolactin levels are normal in obese subjects(81). However, prolactin responses to standard stimuli including insulin, thyrotropin releasing hormone,and dopamine blockers are subnormal in obese patients (82). The responses are improved by weight loss (83), but some abnormalities may persist such as the response to insulin-induced hypoglycemia(84) suggesting some underlying hypothalamic abnormality. The significance of this abnormality is not established. However, prolactin has been proposed as an important hormone for the development of breast cancer having been shown to affect mammary epithelial cell function.

HYPERPARATHYROIDISM

Obese patients have elevated levels of parathyroid hormone compared to lean subjects (85) and the elevation correlates with degree of obesity (86). Primary hyperparathyroidism due to a parathyroid adenoma or parathyroid gland hyperplasia is associated with both elevated serum calcium levels and elevated levels of parathyroid hormone. In obese patients where calcium levels are normal or low secondary hyperparathyroidism occurs by definition. In this situation, the elevated parathyroid hormone levels are presumed to be secondary to a primary deficiency of body stores of calcium or some other regulatory hormone such as vitamin D. The cause of the elevated parathyroid hormone levels in obese patients remains unknown as is the significance of this observation. However, vitamin D analogs have been shown to have differentiating effects in several model systems including HL-60, human breast cancer cell lines, and human prostate cancer cell lines. In addition, some epidemiologists have found that rates of common forms of cancer are lower near the equator and have suggested that this observation relates to the increased synthesis of vitamin D from 7-dehydrocholesterol in skin resulting from exposure to sunlight.

DECREASED SERUM TESTOSTERONE

Obese men have been shown to have normal libido, secondary sexual characteristics, potency, and testicular size in spite of low or borderline low levels of total serum testosterone (87). This discrepancy is explained by the fact that approximately 98 percent of circulating testosterone is bound to serum proteins, primarily sex hormone binding globulin (SHBG). In obese men, SHBG levels are reduced leading to a reduced total serum testosterone level with normal free levels of testosterone (88). Since free testosterone is the biologically active form, there are no associated abnormalities of reproductive hormonal function in obese men without other reproductive disorders. The changes found in serum testosterone and estrogen may impact prostate cancer as discussed in the chapter on prostate cancer. To the extent that changes in sex hormone profiles may occur secondary to obesity and high fat diets, these considerations are relevant to the etiology of prostate cancer.

CONCLUSION

Clearly, the endocrine system plays an integral role in the etiology and maintenance of the obese state and in the etiology of many common forms of cancer. Many of the changes observed in the obese patient are epiphenomena of unknown significance, but others may be important in cancer development, promotion, or progression. Abnormalities in insulin secretion and action are central to many of the observed metabolic changes in the obese patient, and may play a role in the etiology and maintenance of the obese state as well as associated forms of cancer. Similarly, reproductive hormonal abnormalities appear to be closely related to obesity and may have special significance for understanding obesity in women throughout the various stages of development and during pregnancy. Furthermore, the development of breast cancer, uterine cancer and ovarian cancer in women may be related to the changes occurring in reproductive hormones during different stages of the reproductive life cycle. In designing strategies to address the problem of common forms of cancer in the U.S. population and in the world, an understanding of the role of the endocrine system in obesity should be useful for both prevention and treatment.

REFERENCES

1. Garrow JS. Energy balance and obesity in man. 2nd edition. Elsevier/North Holland Inc., New York, 1978, p. 243.
2. Bray GA. Obesity: a clinical and metabolic approach. Med. Clin. N. America 1989; 73:1-269.
3. Brownell KD, Foreyt JP (eds). Handbook of Eating Disorders. Physiology, Psychology, and Treatment of Obesity, Anorexia and Bulimia. Basic Books, New York.
4. Stunkard AJ, Sorensen TIA, Harris C. Adoption study of human obesity. New Engl J. Med. 1986; 314:193-198.

5. Heber D. Endocrinology of Obesity. In: Obesity: Pathophysiology, Psychology and Treatment. (G.Blackburn and B.S. Kanders, eds.) Andover Medical Publishers, Boston, 1994 pp. 98-110.
6. Heber D. Endocrine responses to starvation, malnutrition, and illness. In: DeGroot LJ, Besser M, Burger HG, Jameson LJ, Loriaux DL, Marshall JC, Odell WD, Potts JT, Rubenstein AH, eds. Endocrinology (3rd edition),W.B. Saunders, Philadelphia, 1994 pp. 2263-2678.
7. Heber D.The role of nutrition in cancer prevention and control.Oncology 1992; 6: 9-14.
8. Cloninger CR, Rice J, and Reich T. Multifactorial inheritance with cultural transmission and assortative mating II. A general model of combined polygenic and cultural inheritance. Am J Hum Genet. 1979; 31:176-198.
9. Chandra RK. The nutrition-immunity-infection nexis: The enumeration and functional assessment of lymphocyte subsets in nutritional deficiency. Nutr Res 1983; 3:605-615.
10. Bennett PH, Rushforth NB, Miller M, and Lecompte P. Epidemiologic studies of diabetes in the Pima Indians. Rec. Prog. Horm. Res. 1978; 108:497-452.
11. West, KM Epidemiology of Diabetes and its Vascular Complications. New York Elsevier 1978, p231.
12. West KM, Kalbfleisch, JM Influence of nutritional factors on prevalence of diabetes. Diabetes 1971; 20: 99.
13. Bagdade, JD, Bierman EL, Porte, D The significance of basal insulin levels in the evaluation of the insulin response to glucose in diabetic and nondiabetic subjects. J Clin Invest 1967; 46:1549.
14. Faber OK, Christensen, K, Kehlet, H et al Decreased insulin removal contributes to hyperinsulinemia in obesity. J Clin Endo Metab 1981; 53, 618.
15. Howard BV, Klimes I, Vasquez B et. al. The antilipolytic action of insulin in obese subjects with resistance to its glucoregulatory action. J Clin Endo Metab 1984; 58:544.
16. Ong JM, Kirchgessner TG, Schotz MC, Kern PA Insulin increases the synthetic rate and messenger RNA level of lipoprotein lipase in isolated rat adipocytes. J.Biol.Chem. 1988;263: 129-33.
17. Rodin J Insulin levels, hunger, and food intake: An example of feedback loops in body weight regulation. Health Psychol 1985; 4:1-18.
18. Hartz, AJ, Barboriak, PH, Wong A et.al. The association of obesity with infertility and related menstrual abnormalities in women. Int J Obes 1987; 3:57.
19. Nimrod A, Ryan KH: Aromatization of androgens by human abdominal and breast fat tissue. J Clin Endo Metab 1975; 40:367.
20. Vague, J The degree of masculine differentiation of the obesities: A factor determining predisposition to diabetes, atherosclerosis, gout, and uric calculous disease. Am J Clin Nutr 1956; 4:20.
21. Kissebah AH, Evans DJ, Peiris A, et al Endocrine characteristics in regional obesities: Role of sex steroids. In Vague J, Bjorntorp P, Guy-Grand B et al (eds).: Metabolic Complications of Human Obesities. Amsterdam, Excerpta Medica,1985, p. 115.
22. Spain DM, Nathan DJ, Gellis M Weight, body type, and the prevalence of coronary atherosclerotic heart disease in males.Am J Med Sci 1963; 245: 63.
23. Rebuffe-Scrive M, Eldh J, Hafstrom LO et al Metabolism of mammary, abdominal, and femoral adipocytes in women before and after the menopause. Metabolism 1986; 35: 792-797. 24. Kerr MG: Significance of weight gain in pregnancy. Lancet 1969; 1: 663.
25. Rebuffe-Scrive M Metabolic differences in fat depots. In Bouchard C, Johnston FE (eds) Fat distribution during growth and later health outcomes. New York Alan R Liss 1988 p. 175.
26. Rebuffe-Scrive M, Enk L, Crona N et al Fat cell metabolism in different regions in women. J Clin Invest 1985; 75: 1973-1976.
27. Leibel RH, Hirsch J Site and sex-related differences in adrenoreceptor status of human adipose tissue. J Clin Endo Metab 1987; 64: 1205-1210.
28. Lyons, PM, Truswell AS, Mira M, et al Reduction of food intake in the ovulatory phase of the menstrual cycle. Am J Clin Nutr 1989; 49: 1164-1168.
29. Tchekmedyian NS, Hickman M, and Heber D Treatment of anorexia and weight loss in patients with cancer or acquired immunodeficiency syndrome. Sem in Oncol 1991; 18:35-42.
30. Pike MC, Krailo MD, Henderson BE, et al. "Hormonal" risk factors, "breast tissue age", and the age-incidence of breast cancer. Nature 1983; 303:767-770.
31. Pike MC, Spicer DV, Dahmoush L, et al. Estrogens, progestogens, normal breast cell proliferation and breast cancer risk. Epidemiol Rev 1993; 15:17-35.
32. Kelsey JL. A review of the epidemiology of human breast cancer. Epid Rev 1979; 1: 74-109.
33. Prentice R, Thompson D, Clifford C, Gorbach S, Goldin B, Byar D. Dietary fat reduction and plasma estradiol concentration in healthy postmenopausal women. JNCI 1990;82:129-134.
34. Heber D, Ashley JM, McCarthy WJ, Chang C-J, Elashoff R. Assessment of adherence to a low fat diet for breast cancer prevention. Prev Med 1992; 21:218-227.

35. Bagga D, Ashley JM, Geffrey S, Barnard JA, Wang H-J, Korenman S, Heber D. Effects of very low fat high fiber diet on reproductive hormones:Implications for breast cancer prevention. Cancer 1995; 22: 351-355.
36. Reichman ME, Judd JT, Longcope C, Schatzkin A, Clevidence BA, Nair PP, Campbell WS, Taylor PR. Effects of alcohol consumption on plasma and urinary hormone concentrations in premenopausal women. JNCI 1993; 85:722-726.
37. Grodin JM, Siiteri PK, MacDonald PC. Source of estrogen production in postmenopausal women. J Clin Endocrinol Metab 36:207-210, 1973.
38. Siiteri PK, MacDonald PC. Role of extraglandular estrogen in human endocrinology. In: Greep RO, Astwood EB, eds. Handbook of Physiology Sec.7, vol.2, part 1 Wash. D.C. Am Physiol Soc 1973;15-629.
39. Schindler AE, Ebert A, Friedrich E. Conversion of androstenedione to estrone by human fat tissue. J Clin Endo Metab 1972; 35:627-630.
40. Simpson ER, Ackerman GE, Smith ME. Estrogen formation in stromal cells of adipose tissue of women: induction by glucocorticoids. Proc Nat Acad Sci 1981; 73:5690-5694.
41. Zumoff B. Relationship of obesity to blood estrogens. Cancer Res.(suppl) 1982; 42:3289-3294.
42. Kalish L. Relationship of body size with breast cancer. J Clin Oncol 1984; 2:287-293.
43. Abe R, Kumagai N, Kimura M. Biological characteristics of breast cancer in obesity. Tohoku J Exp Med 1976;120:351-9.
44. Donegan WL, Hartz AJ, Rimm AA. The association of body weight with recurrent cancer of the breast. Cancer 1978; 41:1590-1594.
45. Howson CP, Kinne D, Wynder EL. Body weight, serum cholesterol and stage of primary breast cancer. Cancer 1986; 58:2372-2381.
46. deWaard F. Epidemiology of breast cancer: a review. Eur J Clin Oncol 1983;19:1671-1676.
47. Verreault R, Brisson J, Deschenes L, Naud F. Body weight and prognostic indicators in breast cancer. Am J Epidem 1989; 129: 260-268.
48. Camoriano JK, Loprinzi CL, Ingle JN, Therneau TM, Krook JE, Veeder MH Weight change in women with adjuvant therapy or observed following mastectomy for node-positive breast cancer. J Clin Onc 1990; 8: 1327-1334.
49. Tretli S Height and weight in relation to breast cancer morbidity and mortality. A prospective study of 570,000 women in Norway. Int J Cancer 1989; 44: 23.
50. Newman SC, Miller AB, and Howe GR A study of the effect of weight and dietary fat on breast cancer survival time. Am J Epidemiol 1986; 123: 767.
51. deWaard F, Poortman J, Collette BJ Relationship of weight to promotion of breast cancer after menopause. Nutr Cancer 1981; 2: 237-240.
52. McNee RK, Mason BH, Neave LM et al Influence of height, weight,and obesity on breast cancer incidence and recurrence in Auckland,New Zealand Breast Cancer Res Treat 1987; 9:145.
53. Eberlein I, Simon R , Fisher S et al Height, weight, and risk of breast cancer relapse. Breast Cancer Res Treat 1985; 5:81.
54. Hebert JR, Augustine A , Barone J, Kabat GC, Kinne DW, Wynder EL. Weight, height, and body mass index in the prognosis of breast cancer: early results of a prospective analysis. 1988; 42:315.
55. Lubin F, Ruder AM, Wax Y, Modan B. Overweight and changes in weight throughout adult life in breast cancer etiology. A case-control study. Am J Epidemiol 1985; 122:579.
56. Le Marchand, L , Kolonel LN, Earle ME, Mi MP. Body size at different periods of life and breast cancer risk. Am J Epidemiol 1988; 128: 137.
57. London SJ, Colditz GA, Stampfer MJ, et al. Prospective study of relative weight, height and risk of breast cancer. JAMA 1989; 262: 2853.
58. Ballard-Barbash R, Schatzkin A, Carter CL, et al Body fat distribution and breast cancer in the Framingham study. J Natl Cancer Inst 1990; 82: 286.
59. Reichman ME, Judd JT, Taylor PR, Nair PP, Jones DY, Campbell WS. Effect of Dietary Fat on Length of the Follicular Phase of the Menstrual Cycle in a Controlled Diet Setting. J Clin Endo Metab 1992; 74:1171-1175.
60. Rose DP, Boyar AP, Cohen C et al. Effect of a low-fat diet on hormone levels in women with cystic breast disease. I. Serum steroids and gonadotropins. J Nat Cancer Inst 1987; 78:623-9..
61. Rose DP, Goldman M, Connolly J, et al. High fiber diet reduces serum estrogen concentrations in premenopausal women. Am J Clin Nutr 1989; 49:1179-1183.
62. Woods MN, Gorbach SL, Longcope C, Goldin BR, et al. Low fat, high fiber diet and serum estrone sulfate in premenopausal women. Am J Clin Nutr 1989; 49:1179-1183.

63. Bernstein L, Yuan J-M, Ross RK, Pike MC, Hanisch R, Lobo R, Stanczyk F, Gao Yu-T, Henderson BE. Serum hormone levels in pre-menopausal Chinese women in Shanghai and white women in Los Angeles: results from two breast cancer case-control studies. Cancer Causes and Control 1990; 1:51-58.
64. Smals AGH, Kloppenberg PWC, Benraad TJ: Body proportions and androgenicity in relation to plasma testosterone levels in Klinefelter's syndrome. Acta Endocrinol 1974; 77:387.
65. Odell WD Isolated deficiencies of anterior pituitary hormones: symptoms and diagnosis. J Am Med Assoc 1966; 197:1006.
66. Vermeulen, A , Verdonck L: Sex hormone concentrations in postmenopausal women. Relation to obesity, fat mass, age, and years postmenopause. Clin Endocrinol 1978; 9: 59.
67. Arendt EC, Pattee CJ: Studies on obesity. I. The insulin-glucose tolerance curve. J Clin Endocrinol 1956; 16: 367.
68. Meistas MT, Foster GV, Margolis S et al: Integrated concentrations of growth hormone, insulin, C-peptide and prolactin in human obesity. Metabolism 1982; 31: 1224.
69. Kopelman PG, Noonan K, Goulton R et al; Impaired growth hormone response to growth hormone releasing factor and insulin in obesity. Clin Endocrinol 1985; 23: 87.
70. Kalucy RS Crisp AH Chart T Nocturnal hormonal profiles in massive obesity, anorexia nervosa and normal females. J Psychosom Res 1976; 20: 595.
71. Beck P, Koumans JHT, Winterling CA et al; Studies of insulin and growth hormone secretion in human obesity. J Lab Clin Med 1964; 64:454.
72. Rabinowitz D, Merimee T, Nelson JK: The hormonal profile in obesity. Trans Assoc Am Physicians 1967; 80:190.
73. Hansen AP: Serum growth hormone repsonse to exercise in nonobese and obese normal subjects. Scand J Clin Lab Invest 1973; 31:175.
74. Londono, JH, Gallagher TF, Bray GA: Effect of weight reduction, triiodothyronine, and diethylstilbestrol on growth hormone in obesity. Metabolism 1969; 18:986.
75. Kelijman M , Frohman LA: Enhanced growth hormone responsiveness to GH - releasing hormone after dietary manipulation in obese and non-obese subjects. J Clin Endo Metab 1988; 66:489.
76. Loche S, Cappa M, Berrelli P et al Reduced growth hormone response to growth hormone releasing hormone in children with simple obesity: Evidence for somatomedin-C mediated inhibition. Clin Endocrinol 1987; 27:145.
77. Sims EAH, Horton ES: Endocrine and metabolic adaptation to obesity and starvation. Am J Clin Nutr 1968; 21:1455.
78. Crockford PM, Salmon PA: Hormones and obesity: changes in insulin and growth hormone secretion following surgically induced weight loss. Can Med Assoc J 1970; 103: 147.
79. Williams T, Berelowitz M, Joffe SN et al Impaired growth hormone responses to growth hormone releasing factor in obesity: a defect reversed with weight reduction. N Engl J Med 1984; 331:1403.
80. Rosskamp R, Becker M, Soetadji S: Circulating somatomedin-C levels and the effect of growth hormone releasing factor on plasma levels of growth hormone and somatostatin-like immunoreactivity in obese children. Eur J Pediatr 1987; 146:48.
81. Wilcox RG Triiodothyronine, TSH, and prolactin in obese women. Lancet 1977; 1: 1027.
82. Cavagnini F, Maraschini C, Pinto M et al: Impaired prolactin secretion in obese patients. J Endocrinol Invest 1981; 4:149.
83. Kopelman PG, White N, Pilkington TRE: Impaired hypothalamic control of prolactin secretion in massive obesity. Lancet 1979; 1:747.
84. Kopelman PG, Pilkington TRE, White N , et al Persistence of defective hypothalamic control of prolactin secretion in some obese women after weight reduction. Br MEd J 1980; 281:358.
85. Atkinson RL, Dahms WT, Bray GA et al: Parathyroid hormone levels in obesity: Effects of intestinal bypass surgery. Min Electr Metab 1978; 1:315.
86. Bell NH, Epstein S, Greene A et al: Evidence for alteration of the vitamin D-endocrine system in obese subjects. J Clin Invest 1985; 76:370.
87. Glass AR, Swerdloff RS, Bray GA, et al: Low serum testosterone and sex hormone binding globulin in massively obese men. J Clin Endocrinol Metab 1977; 45:1211.
88. Amatruda JM, Harman SM, Pourmotabbed G et al:Depressed plasma testosterone and fractional binding of testosterone in obese males. J Clin Endo Metab 1978; 47:268.

3

NUTRITIONAL, HORMONAL, AND ENVIRONMENTAL MECHANISMS IN BREAST TUMORIGENESIS

John Glaspy

Division of Hematology and Oncology
UCLA School of Medicine
STE120, 200 Medical Plaza
Los Angeles, California 90095

Epithelial cancer of the breast is a common malignancy in women in the United States. Although much is known about the risk factors for this disease, the pathophysiology of its development and progression are incompletely understood and effective preventative strategies have not been developed. Modern therapy for breast cancer, consisting of surgery, radiotherapy and systemic therapy with hormonal and chemotherapeutic agents has not been successful in eliminating this disease as a major cause of cancer death in this country. Recent discoveries regarding the molecular biology and genetics of breast cancer promise to advance our understanding of the biology of this disease, and may lead to more effective prevention and treatment strategies. Similarly, a better definition and understanding of the environmental determination of breast cancer risk, and the interaction of these determinants with the molecular biology of epithelial cells in the breast could support the rational development of even more effective and feasible approaches.

Breast cancer incidence rates vary by fivefold between low risk countries such as Japan, and high risk countries, including the United States, Australia and Northern Europe [1]. The offspring of migrants from low to high risk countries have risks of breast cancer close to those ambient in their new country, establishing the environmental basis for these international variations [2-4]. Studies to date have indicated that environmental factors such as radiation exposure, dietary fiber [5,6], alchohol ingestion [7-12], and intake of dietary micronutrients such as beta carotene [13-16], vitamins C and E [14], or selenium [17,18] cannot explain these observations.[1,19]. While the aspect of diet most important in determining breast cancer risk remains both unknown and controversial, most studies have suggested three possible candidates: 1) total dietary fat and calories, 2) the composition of unsaturated fat in the diet and 3) dietary content of estrogenic substances, including phytoestrogens and organochlorine compounds. The available data do not support the preponderance of any of these factors over the others in promoting the high risk of breast cancer in the United States and other high risk countries. An understanding of the relative importance of each factor and, ultimately, the mechanism by which it alters the biology of breast epithelial cells will

Dietary Fats, Lipids, Hormones, and Tumorigenesis, edited by Heber and Kritchevsky
Plenum Press, New York, 1996

be essential to the rational integration of dietary manipulation into modern prevention and treatment schemes.

TOTAL DIETARY FAT AND CALORIES AND BREAST CANCER RISK

Because of the high caloric density of dietary fat, it has been difficult to isolate the effect of dietary fat from total caloric intake, and many of the animal studies of dietary fat and breast cancer have not been designed to study fat intake independent of total energy consumption [20]. Case control studies have found an increase in fat intake in breast cancer patients [21-23], although pooled data from all studies suggest only a modest relative risk associated with large differences in dietary fat. A study of bias in food reporting suggests that the differences observed in case control studies is artifactual [24]. Prospective cohort studies of up to ten years have not detected an increased risk of breast cancer with higher fat intakes [1,5,25,26]. Population studies have not observed a strong association of dietary fat intake with the incidence of breast cancer [27]. Finally, although the incidence of breast cancer in the United States has been stable or increasing, the intake of dietary fat has been decreasing [28]. While it remains possible that very low fat intakes (less that 15 gm/d) may confer a protection from breast cancer which has not been detected in these studies in which no group had sufficiently low fat consumption, this seems unlikely. It is more plausible that fat intake exerts its effects early in life, or that dietary fat is complex, with components which decrease or increase risk. If these components of dietary fat are not known and controlled for, the impact of dietary fat in any one study may increase, decrease or leave unchanged the risk of breast cancer. The effect of dietary fat on the incidence of breast cancer in the United States is currently under study in the prospective randomized intervention Women's Health Trial.

Most proponents of the link between total dietary fat and breast cancer postulate that fat and calories impact on breast cancer through their effects on estrogen levels. Obese women have increased circulating levels of estradiol. A Western type diet is associated with elevated plasma sex hormone levels and decreased levels of their binding globulin, increasing availability to tissues[29-34]. Although in short term intervention studies of low fat diet, significant decreases in estrogen levels have not been observed [35-37], some diet related changes in hormone levels are observed in adolescence [38], and may alter breast cancer risk in later adulthood. Interventional studies have suggested that decreasing dietary intake of calories, fat or complex carbohydrate by adult women may be associated with decreases in circulating levels of prolactin and increases in circulating levels of binding globulins [39]. If changes in the blood levels of free estrogen, progesterone or prolactin can change a woman's risk of developing breast cancer, these observations suggest that dietary fat restriction and weight reduction still may be effective strategies for breast cancer prevention, mediated through hormonal changes independent of any carcinogenic effects of dietary fat. It is possible that the small changes (on the order of 30% [40]) in estrogen levels achievable through diet might interplay with the fatty acid composition of dietary fat (see below) in determining breast cancer risk.

COMPOSITION OF UNSATURATED DIETARY FAT AND BREAST CANCER RISK

The issue of fat composition has not been fully addressed in epidemiological studies. The results of studies of unsaturated fat intake and breast cancer risk have been mixed; with both increased [41-43] and decreased [5,23] risk associated with unsaturated fat ingestion. These observations are consistent with the hypothesis that some polyunsaturated fatty acids (PUFA) increase the risk of breast cancer and others have a protective effect. In most populations, the bulk of fat ingested is saturated and the PUFA composition of ingested fat cannot be easily studied.

Feeding studies in animals have permitted a focus on the role of dietary fat composition in mammary cancer promotion, particularly on the content of linoleic acid, an omega-6 PUFA (ω-6 PUFA) present in corn and safflower oils. In rodents, diets high in linoleic acid are associated with an increase in 7,12 dimethylbenzanthracene (DMBA) induced mammary cancer promotion [44,45]. High fat diets containing olive oil or coconut oil do not promote tumorigenesis. Diets rich in corn oil stimulate growth and pulmonary metastases of estrogen-independent human breast cancer cell lines growing in athymic mice [46-48]. These *in vivo* observations have been supported by *in vitro* studies in which linoleate has been found to stimulate the growth of breast cancer cell lines cultured in serum-free medium [49,50]. The hypothesis that ω-6 PUFA may play a role in human breast cancer is supported by the trends in linoleate intake and breast cancer incidence in the U.S. Diet in the United States has shifted from animal fats to substitutes frequently based on corn oil; dietary intake of linoleate has risen, as has the incidence of breast cancer.

A mechanism by which ω-6 PUFA may alter the biology of breast epithelium can be proposed. Linoleic acid (C 18:2, ω-6) is converted to arachidonic acid (C 20:4, ω-6) by desaturation and elongation reactions. Arachidonic acid is then rapidly incorporated into membrane phospholipids by transacylases. Many exogenous stimuli, including growth factors, result in the release of arachidonic acid from membrane phospholipids with subsequent conversion to 2-series prostaglandins by cyclooxygenases [51]. Some eicosanoids, including the 2 series prostaglandins and 4 series thromboxanes, have been associated with mammary cancer promotion and high levels of eicosanoids have been found in mammary tumors in animals and in human breast cancers [52]. The hypothesis that the tumor promotion observed with ω-6 PUFA feeding is related to ω-6 derived 2-series eicosanoid production and an increase in w-6 arachidonic acid in the cell membranes of mammary cells [53] is supported by the observation that indomethacin blocks both DMBA-induced breast tumorigenesis [54], as well as the linoleate enhancement of growth and metastasis of a transplantable mouse mammary tumor [55].

The availability of ω-6 PUFA for its biological action may be influenced by the overall PUFA composition of dietary fat. Competition between classes of PUFA may determine the products of common enzymes involved in their metabolism to eicosanoids. In the DMBA treated rat, dietary ω-3 PUFA, such as eicosapentaenoic acid (EPA; C 20:5, ω-3), or conjugated linoleate [56,57] retards mammary tumor development compared to standard diets when pair fed at equal fat calories in the diet [58]. This inhibitory action of ω-3 PUFA may be mediated by an inhibition of the production of the 2-series in favor of the 3-series and 4-series eicosanoids, which are less inflammatory. If ω-6 PUFA and ω-3 PUFA compete to increase or decrease cell membrane arachidonic acid content and cellular levels of cancer-promoting eicosanoids, it is unlikely that these effects would have been detected in any epidemiological studies. The balance of PUFA constituents in the diet is a viable hypothesis for explaining the role of the environment in determining breast cancer risk. This hypothesis would explain some of the anomalies in international trends in breast cancer incidence. The risk of breast cancer is high in Northern Europe and relatively low in Southern Europe; although the two populations consume diets with similar total fat contents, the use of olive oil, an ω-9 non-inflammatory PUFA, in Southern Europe can be considered as conferring a favorable ω-x/ω-6 PUFA ratio in the unsaturated fat component of the diet. In animal studies, olive oil is protective against the development of breast cancer [59] and epidemiologic studies in humans have found a protective effect of olive oil [60].

DIETARY CONTENT OF ESTROGENIC SUBSTANCES AND BREAST CANCER RISK

Some of the effects of diet on estrogen target tissues may be due to estrogen like substances in the diet [34,61], which may act to enhance or inhibit the effects induced by endogenous hormones. Some of these substances are produced naturally; others are synthetic substances to which people in industrialized countries are exposed. Much attention has been focused on the phytoestrogens, a group of plant derived compounds which are present in large amounts in soy products [62]. These compounds compete with estrogen for binding to its receptor [63], and as weak agonists of estrogen may exert an estrogen blocking effect [34]. The phytoestrogens may also increase the rate of metabolism of endogenous estrogen [64], decrease the rate of estrogen production by inhibiting aromatases[65] and increase the level of binding globulin [66]. High levels of these phytoestrogens are found in the urine of Japanese subjects consuming a traditional diet and it has been suggested that intake of these compounds accounts in part for the low incidence of breast and prostate cancer in Japan [67]. Recent interest has focused on the phytoestrogen genistein, which is found in large amounts in tofu and has both weak estrogen agonist effects and potent tyrosine kinase inhibiting activity [68]. Genistein inhibits the proliferation of human breast cancer cells *in vitro* [69,70], and may inhibit oncogene expression in murine cells [71] and tumor angiogenesis [72]. Phytoestrogen intake may be protective with respect to breast cancer by altering the availability of endogenous estrogens for receptor interaction and by inhibiting tyrosine kinase which is essential to the signal transduction of many growth factors and relevant oncogenes, including HER-2/neu[73].

Synthetic estrogenic substances are produced in the manufacture of organochlorine pesticides and are present in the drinking water and concentrated in the plant and animal fat present in the diet of industrialized countries[74]. The best studied of these compounds is 2,2-bis(p-chlorophenyl)-1,1,1-trichloroethane (DDT). Although this compound is no longer produced in the United States, substantial levels remain in the environment. DDT has potent estrogenic activity [75]. Although efforts to correlate organochlorine exposure with breast cancer risk in case control studies have produced inconsistent results [76], most of these studies have used levels in stored blood of organochlorines and their metabolites. This approach to the study of the impact of organochlorines on the development of breast cancer is appropriate if these agents are acting as chemical carcinogens, producing damage that results in malignancy years later. If these agents are acting as promoters (strong estrogens), exposure should be measured at a time point closer to clinical presentation [77]. Moreover, because these agents are fat soluble, mammary fat levels of the compounds may more accurately reflect the relevant exposure. In a case control study that examined the levels of organochlorines in mammary fat, patients with breast cancer, especially those with estrogen receptor positive tumors, were found to have significantly higher levels [78].

NUTRITIONAL EFFECTS ON THE PROGRESSION OF ESTABLISHED BREAST CANCER

Nutrition may also modulate the natural history of established breast cancer, especially the progression of micrometastases. In a carcinogen induced mammary carcinoma model in the rat, a 30% reduction in caloric intake was associated with a decreased tumor recurrence rate after tumor removal, without changing circulating estrogen or testosterone levels or adversely affecting normal growth [79]. Weight gain is common in women undergoing adjuvant chemotherapy treatment for breast cancer. It has been found that the greater this weight gain, the poorer the prognosis in terms of systemic relapse [80]. In a study of disease-free survival at 10 years of 923 post-operative breast cancer patients, obesity at diagnosis (>125% of optimal body weight) was associated with a greater risk for recurrence (42%) compared to non-obese patients (32%) [81]. In the patients without lymph node involvement, the difference was even greater (32% vs. 19%). The same processes which mediate the breast carcinogenesis associated with Western diets may be responsible for the aggressive behavior of established breast cancer observed in association with obesity.

SUMMARY AND SUPPORTIVE EXPERIMENTAL DATA

Our paradigm for understanding the effects of diet on the development and clinical progression of established breast cancer can be shown schematically:

In vitro Data: Interaction of PUFA and Significant Molecular Alterations

In attempting to study the interaction of 30% of human breast cancers overexpress the HER-2/neu gene and this subset of tumors exhibits specific clinical behavior with a shorter associated disease free and overall survival [73]. Overexpression of this gene serves as a prognostic factor in breast cancer. Two possible explanations for this are: a) expression of this gene is a marker of unfavorable outcome but plays no role in its pathogenesis, a useful epiphenomenon, or b) the gene plays a role in pathogenesis. Circumstantial data support the latter hypothesis, however, much of these data were derived using a mutated rat HER-2/neu gene [82], or a human gene in murine fibroblast cells [83,84]. To address the question, investigators in our Division cloned and sequenced the human gene from a breast malignancy and placed the gene in a retroviral expression vector. Multiple copies of the gene were then introduced into the human breast cancer cell line MCF-7. which is known to contain a single copy of the gene and express low levels of the gene product. This converted the single-copy, low expressing human breast cancer cell line to a multiple copy, high expressing cell line. The levels of overexpression selected were the same as those seen in human breast cancers. After generating this cell line, termed MCF-7/10, studies were undertaken to characterize

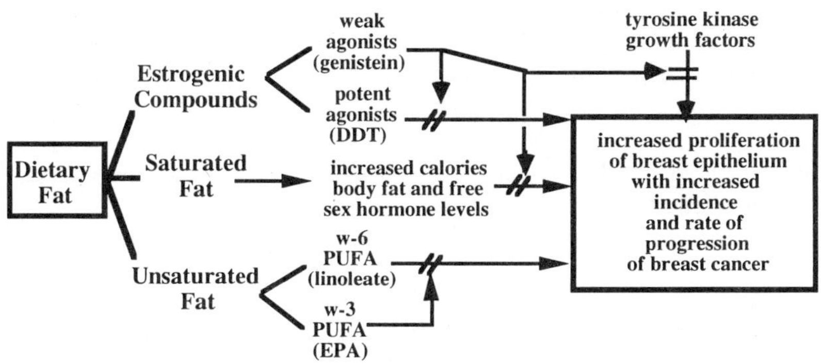

Figure 1. In ongoing studies, we have begun to explore the validity of this paradigm using *in vitro* and *in vivo* systems, as well as the initiation of human clinical trials.

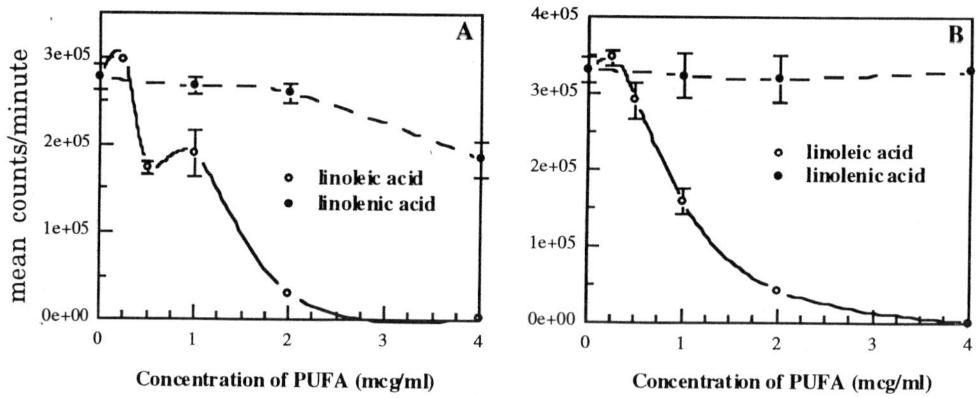

Figure 2. Thymidine incorporation of MCF-7 (A) and MCF-7/10 cells (B) cultured in the presence of linoleic acid (ω-6) or linolenic acid (ω-3). Error bars represent standard errors (6 wells per data point).

any biologic changes the overexpression of HER-2/neu mediated. MCF-7/10 cells with moderate and high levels of HER-2/neu overexpression were compared to MCF-7 cells transfected in an identical fashion with a vector lacking the HER-2/neu c-DNA as well as the parental MCF-7 line. These studies demonstrated that the HER-2/neu overexpressing cells have significantly higher rates of DNA synthesis, as measured by tritiated thymidine incorporation, compared to the control parental cell line and significantly increased *in vitro* growth rate as measured by cell counts. In the soft agar cloning assay, which serves as an *in vitro* measure of anchorage independent growth, the MCF-7/10 cells have a cloning efficiency 3 times that of the control MCF-7 cells. MCF-7/10 cells were able to form tumors in the nude mouse at a much greater rate than the MCF-7 control and the MCF-7/10 derived tumors grew much faster. MCF-7 is an estrogen dependent line; when engineered to MCF-7/10, the cells grow in the nude mouse without estrogen supplementation, although at a slower rate than in the presence of estrogen. The MCF-7 and MCF-7/10 lines are two human breast cancer cell lines which differ only in the introduction of a clinically relevant molecular alteration which confers a "more malignant" phenotype.

We have done preliminary experiments using a modification of published techniques [49]. The effects of PUFA on the rate of DNA synthesis by MCF-7 or MCF-7/10 cells was measured by their incorporation of ^{3}H-thymidine. 5 x 10^3 cells per well were plated in 96 well plates (Millipore) in RPMI medium containing 1% calf serum and no phenol red which had been charcoal extracted to remove estrogenic compounds. After 24 hours growth at 37°C in a 5% CO_2 atmosphere, ω-6 PUFA, ω-3 PUFA or combinations of PUFAs in various concentrations were added to triplicate wells. The cells were incubated for another 48 hours and then received a 2 μCi pulse of ^{3}H-thymidine. Thymidine incorporation was stopped after 4 hours of incubation by filtration and washing of wells with PBS. DNA was precipitated with 5% ice-cold TCA, .2 μL/well for 5 minutes and air dried. ^{3}H-thymidine was measured in a scintillation counter.

In vitro, linoleic acid is stimulatory to breast cancer cells at lower concentrations (.25-.75 μg/ml). We have not found stimulation or suppression of proliferation from linolenic acid or EPA. We have not found differences in the proliferative response to linoleic acid between the MCF-7 and MCF-7/10 cells. Mixture experiments to study the effects of ω-3 PUFA on the proliferative response of breast cancer cells to low concentration of ω-6 PUFA have been carried out. In thymidine incorporation assays using MCF-7 cells, the mean counts per well were 189,866 in wells containing linoleic acid alone; in wells containing the same

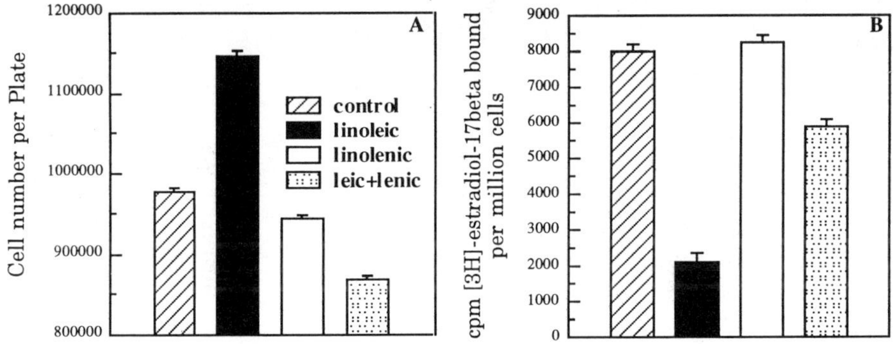

Figure 3. A: MCF-7/10 tumor cell counts after seven days of culture in an estrogen free medium with medium alone, linoleic acid (1 μg/ml), linolenic acid (1 μg/ml), or linoleic acid (1 μg/ml) with linolenic acid at an 8:1 molar ratio (linoleate:linolenate). B: Estradiol binding to MCF-7/10 cells from panel A.

concentration of linoleic acid with linolenic acid added in a molar ratio of 16:1 (ω-6:ω-3), mean counts were 51,785 ($p < .01$, Student's t-test), similar to the counts in the control wells (no PUFA). Similar blocking effects of ω-3 PUFA were observed in experiments using MCF-7/10 cells.

In longer term culture, we plated 2.5 x 10^5 MCF-7/10 cells into T-75 wells with estrogen free medium. Cells were cultured for 7 days in the presence of either: medium alone (control), lenoleic acid, 1μg/ml, linolenic acid, 1 μg/ml or the combination of linoleic acid and linolenic acid in an 8:1 molar ratio, with the linoleic acid concentration remaining at 1μg/ml. After seven days of culture at 37^0C and 5% CO_2, the cells were counted manually. After seven days of culture, estrogen receptor analysis of these cells was done by H^3estradiol binding assays.

The results show that the presence of linoleate was associated with increased cell numbers after seven days; this effect was blocked by the presence of linolenic acid. The data indicate that linoleate may be selectively stimulatory to the estrogen receptor negative breast cell clones.

We have also studied the effects of DDT on the MCF-7 and MCF-7/10 cells. MCF-7 or MCF-7/10 cells were cultured in estrogen free RPMI containing 1% calf serum, with and without the addition of DDT at a concentration of 100 nmol. After 48 hours, 2.5 x 10^4 cells per 6 cm dish were plated for soft agar colony assay. In this assay, the cells are plated between a .4% agar bottom layer and a .2% agar top layer. The agar stocks were made with or without DDT. The plates were then incubated at 37° C in 5% CO_2 for three weeks, at which time colonies were counted. Each of the four experiments were repeated in triplicate. The mean colony counts are shown:

	Control	DDT	DDT (% control)
MCF-7:	79 ± 4	95 ± 5	120 ± 6
MCF-7/10:	126 ± 5	184 ± 6	146 ± 4

The colony numbers are low because of the absence of estrogen. The increase in colonies noted with DDT is consistent with its estrogenic effects. HER-2/neu overexpressing cells demonstrated a greater increment in colonies with DDT exposure than the parent cell line. This suggests that DDT may provide a growth advantage to cells with a clinically relevant genetic alteration.

IN VIVO DATA: INTERPLAY OF DIETARY FACTORS AND THE GROWTH OF TUMORS DERIVED FROM HUMAN BREAST CANCER CELLS

We have carried out preliminary *in vivo* studies of the interplay of total calories, PUFA composition and clinically relevant alterations in oncogene expression in the growth of tumors derived from human breast cancer cell lines. In one study, 60 nude mice were implanted with estrogen pellets and injected subcutaneously with 5 X 10^7 MCF-7/10 cells in each of two sites. The injected mice were cared for on standard ad lib diets until visible tumors developed. The 48 surviving mice were then weighed, stratified and randomized to one of four dietary groups:

Diet 1: Ad lib feeding, 20% fat diet, fat as corn oil
Diet 2: Pair fed to animals receiving Diet 1, 20% fat diet, 17.5% fish oil, 2.5% corn oil
Diet 3: Calorie restricted to 70% pair fed to animals receiving Diet 1, 20% fat, corn oil

Group 4 received ad lib feeding with a diet containing 20% of calories as fat, with corn oil as the source of fat (Diet 1, research plan). Group 2 received isocaloric pair feeding to group 4, with a diet containing 20% calories as fat, with 95% of the fat calories from menhaden oil (fish oil) and 5% from corn oil (Diet 2, below). Group 3 was calorie restricted to 70% of the calories consumed by group 4 and received the high corn oil diet fed group 4 (Diet 3, below). Calories consumed were recorded daily, and tumor measurements were taken twice weekly. Tumors in group 3 (calorie restricted) grew significantly more slowly than group 4 and tumors in group 2 (high fat, menhaden oil diet) evidenced a significantly lower growth rate than those in group 4. This suggests a possible inhibitory effect of fish oil, independent of fat or calories.

In this experiment, tumors removed from the mice at the time of sacrificed were studied for HER-2/neu expression using immunohistochemistry. The MCF-7/10 cells injected into the mice at the beginning of the experiment score as 3+ in this assay. All of the tumors removed were positive for HER-2/neu and were scored as either 2+ or 3+. The proportion of tumors that were scored as 2+ were: Group 4=.16, Group 3=.29, and Group 2=.29. These data suggest that there may be little or no effect of these diets on HER-2/neu expression, but further studies will be necessary.

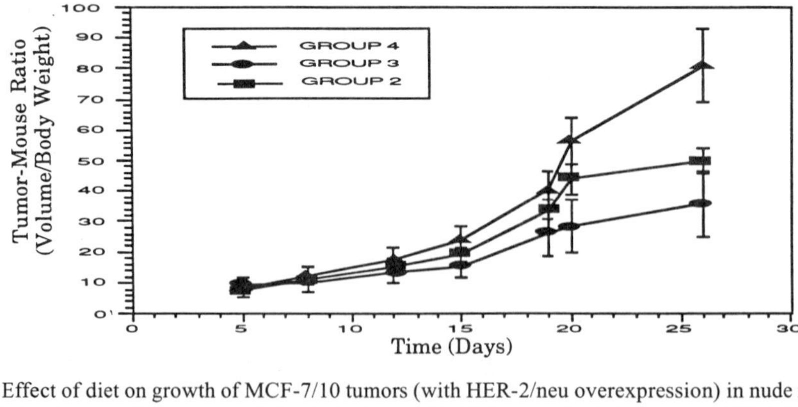

Figure 4. Effect of diet on growth of MCF-7/10 tumors (with HER-2/neu overexpression) in nude mice. Data are presented as tumor volume ratios relative to body weight. Diet groups included: group 2 (high fat, fish oil); group 3 (high fat corn oil, calorie restricted); group 4 (low fat, corn oil) Tumors in group 3 and those in group 2 were both significantly different from group 4 ($P<0.05$).

FUTURE STUDIES

The studies reported above are ongoing and continuing. In the *in vitro* work, we will carry out studies using additional malignant cell lines, as well as additional human cells lines representing normal, immortalized but not transformed and fully malignant phenotypes, with and without overexpression of the HER-2/neu gene. Theses studies will be aimed at understanding the effects of linoleate (ω-6 PUFA), eicosapentanoic (EPA) and linolenic acid (ω-3 PUFAs), estrogen, genistein and DDT on the transformation and growth of human breast epithelial cells. They will also attempt to define the ω-6 to ω-3 ratios, estrogen and DDT concentrations at which inhibition of transformation and growth take place, to serve as a target and an intermediate marker for *in vivo* studies and human interventional trials. The use of cell lines representing normal, pre-malignant and malignant breast epithelium will permit us to study the point(s) in the process by which breast epithelial cells become malignant that PUFA and estrogenic compounds are likely to exert their greatest influence, and to draw conclusions regarding both the interplay of these factors with the significant molecular alterations and the potential success of various nutritional interventions on the development breast cancer as opposed to the treatment of established disease. The results of these experiments will be relevant to the selection of intervention strategies, and the identification of intermediate markers of the efficacy of intervention. For example, it may be necessary to select an intermediate marker that includes a target ω-6 to ω-3 ratio in body fat which is different for premenopausal and postmenopausal women. At the completion of these studies we will have some understanding of the effects of ω-6 and ω-3 PUFA, estrogen, phytoestrogens, and organochlorines on the growth, in three separate assays, of human breast epithelial cell lines representing benign, early progression and malignant phenotypes. If the effects are similar in all cell lines, this would suggest that the mechanisms operative in the nutritional enhancement of growth of established breast cancers may be similar to the mechanisms involved in the promotion of breast cancer development. We will have data on the cumulative effects of, and interplay between, these factors and overexpression of HER-2/neu.

There are theoretical advantages to animal models which utilize human tumors with defined and clinically relevant genetic alterations growing *in vivo*. The effects and interplay of dietary fat, PUFA composition, caloric intake, estrogen levels, phytoestrogen intake, DDT exposure and oncogene expression on the implantation in mammary tissue and subcutaneous growth of human breast tumor cells in nude mice will be addressed in these future studies. A series of pair feeding experiments are being carried out to further define the effects of absolute level of dietary linoleate and ω-3 PUFA, calorie restriction, dietary genistein and DDT exposure on both implantation efficiency and growth rate of tumors derived from various human breast cancer cell lines, with and without overexpression of HER-2/neu, growing in nude mice. At the completion of each nude mouse experiment, the mice will be dissected and samples of plasma, tumor tissue, and body fat will be obtained. Prostaglandin concentrations will be measured in body fat and tumor tissue in all experiments. Samples of tumor, and body fat will undergo fatty acid analysis. Plasma will be assayed for: estrogen, cholesterol, and quantitative fatty acid profile. All tumor samples will be studied for HER-2/neu expression and estrogen receptor by immunoperoxidase staining, as well as for total prostaglandin content (PGF2-a and PGE2) and quantitative fatty acid analysis.

We are also carrying out studies on banked human tissue, including fatty acid and organochlorine pesticide analyses of mammary fat from women with breast cancer and control women. These data will be important for testing hypotheses generated to explain the results of the *in vitro* and nude mouse studies, and building a more complete understanding of the environmental modulation of breast cancer incidence and progression. By examining both fatty acid composition and organochlorine levels, it will be possible to identify the components of dietary fat that are most associated with the development of breast cancer in women from different ethnic groups. For the breast cancer cases, tumor estrogen and progesterone receptor, flow cytometry and HER-2/neu expression data are known, and tumor DNA, RNA, protein and intact samples suitable for immunohistochemical studies are available for each case. Organochlorines have been studied previously as potential factors in the development of breast cancer. Some case control studies have found higher levels of organochlorine compounds in the archived blood of patients with breast cancer compared with controls although this has not been a consistent finding [76,77,85-91]. The positive studies have usually found increased levels of organochlorine pesticide compounds rather than the PCB organochlorines [86]. In these studies of archived blood, it has been assumed that blood organochlorine levels accurately reflect both total exposure over time and exposure of cells in the microenvironment of the human breast. Organochlorines can be stored in the lipid compartment and this has led to some efforts to measure exposure using subcutaneous fat analyses [92]. Limited information is available on the levels of organochlorine pesticides in the breast fat of women with and without breast cancer, and this data suggests that increased levels of these compounds are found in the breast fat of patients with estrogen receptor positive breast cancer [78,93]. This has not been confirmed in a large number of women [76]. This issue is of particular interest because of the known estrogenic activity of DDT, and the interrelationships we are attempting to characterize between the nutritional factors which may alter breast cancer risk and the level of endogenous and exogenous estrogenic stimulation. Information regarding the levels of organochlorines in breast fat in different groups of women in the United States, as well as levels in women with breast cancer from these groups would be very helpful in interpreting the results of our animal studies and translating our animal studies into effective breast cancer prevention and treatment strategies. When these studies are complete, we hope to understand the magnitude of the organochlorine exposure problem, the strength of its association with breast cancer in all segments of California society, and the effects of organochlorine exposure on the importance of other environmental determinants of human breast cancer.

Finally, we have initiated a large trial involving 150 women, 75 with chemotherapy unresponsive metastatic breast cancer and 75 who are clinically free of disease following bone marrow transplantation. In this study, samples of breast and gluteal fat as well as serum are obtained prior to the initiation of a low fat, diet with substantial soy (genistein) and fish oil supplementation. After three months, the tissue sampling will be repeat ed. The fat samples will be studied for their fatty acid composition. Changes over time in fatty acid levels will be correlated with tumor response and relapse rates. As further data from the *in vitro* and *in vivo* studies as well as this clinical trial becomes available, future interventional trials will be developed aimed at bringing our increased understanding of the environmental determination of human breast cancer to bear on the problem of developing successful prevention and intervention strategies.

REFERENCES

1. Hunter DJ, Willett WC: Diet, body size, and breast cancer. Epidemiol Rev 15:110-32, 1993
2. Staszewski J, Haenszel W: Cancer mortality among Polish born in the United States. J Nat Cancer Inst 35:292-6, 1965
3. McMichael AJ, Giles GG: Cancer in migrants to Australia: extending descriptive epidemiological data. Cancer Res 48:751-6, 1988
4. Kelsey JL, Horn-Ross PL: Breast cancer: magnitude of the problem and descriptive epidemiology. Epidemiol Rev 15:7-16, 1993

5. Willett WC, Hunter DJ, Stampfer MJ, et al: Dietary fat and fiber in relation to risk of breast cancer: an 8 year follow-up. JAMA 268:2037-44, 1992
6. Graham S, Zielezny M, Marshall J, et al: Diet in the epidemiology of postmenopausal breast cancer in the New York State Cohort. Am J Epidemiol 136:1327-37, 1992
7. Longnecker MP, Berlin JA, Orza MJ, et al: A meta-analysis of alcohol consumption in relation to risk of breast cancer. JAMA 260:652-6, 1988
8. Hiatt RA, Bawol RD: Alcoholic beverage consumption and breast cancer incidence. Am J Epidemiol 120:676-83, 1984
9. Friedenreich CM, Howe GR, Miller AB, et al: A cohort study of alcohol consumption and risk of breast cancer. Am J Epidemiol 137:512-20, 1993
10. Young TB: A case-control study of breast cancer and alcohol consumption habits. Cancer 64:552-8, 1989
11. Toniolo P, Riboli E, Protta F, et al: Breast cancer and alcohol consumption: a case-control study in northern Italy. Cancer Res 49:5203-6, 1989
12. Gapstur SM, Potter JD, Sellers TA, et al: Increased risk of breast cancer with alcohol consumption in postmenopausal women. Am J Epidemiol 136:1221-31, 1992
13. Marubini E, Decarli A, Costa A, et al: The relationship of dietary intake and serum levels of retinol and beta-carotene with breast cancer. Results of a case-control study. Cancer 61:173-80, 1988
14. Hunter DJ, Manson JE, Colditz GA, et al: A prospective study of the intake of vitamins C, E, and A and the risk of breast cancer [see comments]. N Engl J Med 329:234-40, 1993
15. Ewertz M, Gill C: Dietary factors and breast-cancer risk in Denmark. Int J Cancer 46:779-84, 1990
16. van 't Veer P, Kolb CM, Verhoef P, et al: Dietary fiber, beta-carotene and breast cancer: results from a case- control study. Int J Cancer 45:825-8, 1990
17. van 't Veer P, van der Wielen RP, Kok FJ, et al: Selenium in diet, blood, and toenails in relation to breast cancer: a case-control study. Am J Epidemiol 131:987-94, 1990
18. Hunter DJ, Morris JS, Stampfer MJ, et al: A prospective study of selenium status and breast cancer risk [see comments]. Jama 264:1128-31, 1990
19. Rohan TE, Bain CJ: Diet in the etiology of breast cancer. Epidemiol Rev 9:120-45, 1987
20. Albanes D: Total calories, body weight, and tumor incidence in mice. Cancer Res 47:1987-92, 1987
21. Graham S, Marshall J, Mettlin C, et al: Diet in the epidemiology of breast cancer. Am J Epidemiol 116:68-75, 1982
22. Howe GR, Hirohata T, Hislop TG, et al: Dietary factors and risk of breast cancer: combined analysis of 12 case-control studies. J Natl Cancer Inst 82:561-9, 1990
23. Howe GR, Hirohata T, Hislop TG, et al: Dietary factors and risk of breast cancer: combined analysis of 12 case-control studies [see comments]. J Natl Cancer Inst 82:561-9, 1990
24. Giovannucci E, Stampfer MJ, Colditz GA, et al: A comparison of prospective and retrospective assessments of diet in the study of breast cancer. Am J Epidemiol 137:502-11, 1993
25. van den Brandt PA, van't Veer P, Goldbohm RA, et al: A prospective cohort study on dietary fat and breast cancer risk. Cancer Res 53:75-82, 1993
26. van den Brandt PA, van't Veer P, Goldbohm RA, et al: A prospective cohort study on dietary fat and the risk of postmenopausal breast cancer. Cancer Res 53:75-82, 1993
27. Marshall JR, Qu Y, Chen J, et al: Additional ecological evidence: lipids and breast cancer mortality among women aged 55 and over in China. Eur J Cancer 28A:1720-7, 1992
28. Stephen AM, Wald NJ: Trends in individual consumption of dietary fat in the United States, 1920-1984. Am J Clin Nutr 1990:457-69, 1990
29. Adlercreutz H, Hamalainen E, Gorbach SL, et al: Diet and plasma androgens in postmenopausal vegetarian and omnivorous women and postmenopausal women with breast cancer. Am J Clin Nutr 49:433-42, 1989
30. Adlercreutz H, Fotsis T, Hockerstedt K, et al: Diet and urinary estrogen profile in premenopausal omnivorous and vegetarian women and in premenopausal women with breast cancer. J Steroid Biochem 34:527-30, 1989
31. Adlercreutz H: Diet, breast cancer, and sex hormone metabolism. Ann N Y Acad Sci 595:281-90, 1990
32. Adlercreutz H: Western diet and Western diseases: some hormonal and biochemical mechanisms and associations. Scand J Clin Lab Invest Suppl 201:3-23, 1990
33. Barbosa JC, Shultz TD, Filley SJ, et al: The relationship among adiposity, diet, and hormone concentrations in vegetarian and nonvegetarian postmenopausal women [see comments]. Am J Clin Nutr 51:798-803, 1990
34. Adlercreutz H, Mousavi Y, Clark J, et al: Dietary phytoestrogens and cancer: *in vitro* and *in vivo* studies. J Steroid Biochem Mol Biol 41:331-7, 1992

35. Fentiman IS, Caleffi M, Wang DY, et al: The binding of blood-borne estrogens in normal vegetarian and omnivorous women and the risk of breast cancer. Nutr Cancer 11:101-6, 1988
36. Ingram D, Bennett F, Wood A: Estradiol binding to plasma proteins after changing to a low-fat diet. Nutr Cancer 12:327-32, 1989
37. Crighton IL, Dowsett M, Hunter M, et al: The effect of a low-fat diet on hormone levels in healthy pre- and postmenopausal women: relevance for breast cancer. Eur J Cancer 12:2024-7, 1992
38. Persky VW, Chatterton RT, Van Horn LV, et al: Hormone levels in vegetarian and nonvegetarian teenage girls: potential implications for breast cancer risk. Cancer Res 52:578-83, 1992
39. Bennett FC, Ingram DM: Diet and female sex hormone concentrations: an intervention study for the type of fat consumed. Am J Clin Nutr 52:808-12, 1990
40. Heber D, Ashley JM, Leaf DA, et al: Reduction of serum estradiol in postmenopausal women given free access to low-fat high-carbohydrate diet. Nutrition 7:137-9; discussion 139-40, 1991
41. Wynder EL, Rose DP, Cohen LA: Diet and breast cancer in causation and therapy. Cancer 58:1804-13, 1986
42. Kakar F, Henderson MM: Diet and breast cancer. Clin Nutr 4:119-30, 1985
43. Hursting SD, Thornquist M, Henderson MM: Types of dietary fat and the incidence of cancer at five sites. Prev Med 19:242-253, 1990
44. Hopkins GJ, Kennedy TG, Carroll KK: Polyunsaturated fatty acids as promoters of mammary carcinogenesis induced in Sprague-Dawley rats by 7,12 didethylbenzantracene. J Natl Cancer Inst 66:517-522, 1981
45. Fischer SM, Conti CJ, Locniskar M, et al: The effect of dietary fat on the rapid development of mammary tumors induced by 7,12-dimethylbenz(a)anthracene in SENCAR mice. Cancer Res 52:662-6, 1992
46. Rose DP, Connolly JM, Meschter CL: Effects of dietary fat on human breast cancer growth and lung metastases in nude mice. J Natl Cancer Inst 83:1491-95, 1991
47. Rose DP, Connolly JM: Effects of dietary linoleic acid on human breast cancer cell growth and metastasis in nude mice (Meeting abstract). Br J Cancer 67:24, 1993
48. Rose DP, Connolly JM, Liu XH: Effects of linoleic acid on the growth and metastasis of two human breast cancer cell lines in nude mice and the invasive capacity of these cell lines *in vitro*. Cancer Res 54:6557-62, 1994
49. Rose DP, Connolly JM: Effects of fatty acids and inhibitors of eicosanoid synthesis on the growth of a human breast cancer cell line in culture. Cancer Res 50:7139-44, 1990
50. Shultz TD, Chew BP, Seaman WR: Differential stimulatory and inhibitory responses of human MCF-7 breast cancer cells to linoleic acid and conjugated linoleic acid in culture. Anticancer Res 12:2143-5, 1992
51. Hansen HS: Dietary essential fatty acids and *in vivo* prostaglandin production in mammals, in Bourne GH (ed): World Review of Nutritional Dietetics. Basel, Karger, 1983, pp 102-34
52. Karmali RA: Prostaglandins and cancer. CA 33:29-39, 1984
53. Rose DP: Diet, hormones, and cancer. Annu Rev Public Health 14:1-17, 1993
54. Carter CA, Milholland RJ, Shea W: Effect of the prostaglandin synthesis inhibitor indomethacin on 7,12 dimethylbenzanthracene-induced mammary tumorigenesis in rats fed different levels of fat. Cancer Res 43:3559-69, 1983
55. Hubbard N, Chapkin RS, Erickson KL: Inhibition of growth and linoleate-enhanced metastasis of a transplantable mouse mammary tumor by indomethacin. Cancer Letter 43:111-120, 1988
56. Ip C, Chin SF, Scimeca JA, et al: Mammary cancer prevention by conjugated dienoic derivative of linoleic acid. Cancer Res 51:6118-24, 1991
57. Ip C, Singh M, Thompson HJ, et al: Conjugated linoleic acid suppresses mammary carcinogenesis and proliferative activity of the mammary gland in the rat. Cancer Res 54:1212-5, 1994
58. Karmali RA, Marsh J, Fuchs C: Effect of omega-3 fatty acids on growth of a rat mammary tumor. J Natl Cancer Inst 73:457-61, 1984
59. Welsch CW: Relationship between dietary fat and experimental mammary gland tumorigenesis: a review and critique. Cancer Res 52(suppl):2040S-8S, 1992
60. Martin-Moreno JM, Willett WC, Gorgojo L, et al: Dietary fat, olive oil intake and breast cancer risk. Int J Cancer 58:774-80, 1994
61. Sathyamoorthy N, Wang TT, Phang JM: Stimulation of pS2 expression by diet-derived compounds. Cancer Res 54:957-61, 1994
62. Dwyer JT, Goldin BR, Saul N, et al: Tofu and soy drinks contain phytoestrogens. J Am Diet Assoc 94:739-43, 1994
63. Verdeal K, Brown RR, Richardson T, et al: Affinity of phytoestrogens for estradiol-binding proteins and effect of coumestrol on growth of 7,12-dimethylbenz[a]anthracene-induced rat mammary tumors. J Natl Cancer Inst 64:285-90, 1980

64. Adlercreutz H, Hockerstedt K, Bannwart C, et al: Effect of dietary components, including lignans and phytoestrogens, on enterohepatic circulation and liver metabolism of estrogens and on sex hormone binding globulin (SHBG). J Steroid Biochem 27:1135-44, 1987
65. Adlercreutz H, Bannwart C, Wahala K, et al: Inhibition of human aromatase by mammalian lignans and isoflavonoid phytoestrogens. J Steroid Biochem Mol Biol 44:147-53, 1993
66. Mousavi Y, Adlercreutz H: Genistein is an effective stimulator of sex hormone-binding globulin production in hepatocarcinoma human liver cancer cells and suppresses. Steroids 58:301-4, 1993
67. Adlercreutz H, Honjo H, Higashi A, et al: Urinary excretion of lignans and isoflavonoid phytoestrogens in Japanese men and women consuming a traditional Japanese diet. Am J Clin Nutr 54:1093-100, 1991
68. Sargeant P, Farndale RW, Sage SO: ADP- and thapsigargin-evoked Ca2+ entry and protein-tyrosine phosphorylation are inhibited by the tyrosine kinase inhibitors genistein and methyl-2,5-dihydroxycinnamate in fura-2-loaded human platelets. J Biol Chem 268:18151-6, 1993
69. Monti E, Sinha BK: Antiproliferative effect of genistein and adriamycin against estrogen- dependent and -independent human breast carcinoma cell lines. Anticancer Res 14:1221-6, 1994
70. Pagliacci MC, Spinozzi F, Migliorati G, et al: Genistein inhibits tumour cell growth *in vitro* but enhances mitochondrial reduction of tetrazolium salts: a further pitfall in the use of the MTT assay for evaluating cell growth and survival [see comments]. Eur J Cancer 11:1573-7, 1993
71. Wei H, Wang Y, Barnes S: Inhibitory effect of genistein on tumor promoter-induced c-fos and c- jun mRNA expression in mouse skin (Meeting abstract). Proc Annu Meet Am Assoc Cancer Res 34:A998, 1993
72. Fotsis T, Pepper M, Adlercreutz H, et al: Genistein, a dietary-derived inhibitor of *in vitro* angiogenesis. Proc Natl Acad Sci U S A 90:2690-4, 1993
73. Slamon DJ, Godolphin W, Jones LA, et al: Studies of the HER-2/neu proto-oncogene in human breast and ovarian cancer. Science 244:707-712, 1989
74. Wolff MS, Landrigan PJ: Environmental estrogens [letter]. Science 266:526-7, 1994
75. Bulger WH, Kupfer D: Estrogenic action of DDT analogs. American Journal of Industrial Medicine 4:163-173, 1983
76. Krieger N, Wolff MS, Hiatt RA, et al: Breast cancer and serum organochlorines: a prospective study among white, black, and asian women. Journal of the National Cancer Institute 86:589-599, 1994
77. MacMahon B: Pesticide residues and breast cancer? Journal of the National Cancer Institute 86:572-573, 1994
78. Dewailly E, Dodin S, Verreault R, et al: High organochlorine body burden in women with estrogen receptor-positive breast cancer. Journal of the National Cancer Institute 86:232-234, 1994
79. Bunk B, Zhu P, Klinga K, et al: Influence of reducing luxury calories in the treatment of experimental mammary carcinoma. Br J Cancer 65:845-51, 1992
80. Lees AW, Jenkins HJ, May CL, et al: Risk factors and 10 year breast cancer survival in northern Alberta. Breast Cancer Res Treat 13:1434-1451, 1989
81. Senie RT, Rosen PP, Rhodes P, et al: Obesity at diagnosis of breast carcinoma influences duration of disease-free survival. Ann Int Med 116:26-32, 1992
82. Muller WJ, Sinn E, Pattengale PK, et al: Single-step induction of mammary adenocarcinoma in transgenic mice bearing the activated c-neu oncogene. Cell 54:105-15, 1988
83. Di Fiore PP, Pierce JH, Kraus MH, et al: erbB-2 is a potent oncogene when overexpressed in NIH/3T3 cells. Science 237:178-82, 1987
84. Hudziak RM, Schlessinger J, Ullrich A: Increased expression of the putative growth factor receptor p185HER2 causes transformation and tumorigenesis of NIH 3T3 cells. Proc Natl Acad Sci U S A 84:7159-63, 1987
85. Savitz DA: Re: Breast cancer and serum organochlorines: a prospective study among white, black, and Asian women [letter; comment]. J Natl Cancer Inst 86:1255-6, 1994
86. Wolff MS, Toniolo PG, Lee EW, et al: Blood levels of organochlorine residues and risk of breast cancer. J Natl Cancer Inst 85:648-52, 1993
87. Shames LS, Munekata MT, Pike MC: Re: Blood levels of organochlorine residues and risk of breast cancer [letter; comment]. J Natl Cancer Inst 86:1642-3, 1994
88. Anderson LF: DDT and breast cancer: the verdict isn't in [news; comment]. J Natl Cancer Inst 86:576-7, 1994
89. Kopans DB: Re: Organochlorines and breast cancer [letter; comment]. J Natl Cancer Inst 86:66, 1994
90. Taubes G: Pesticides and breast cancer: no link? [news]. Science 264:499-500, 1994
91. Longnecker MP, London SJ: Re: Blood levels of organochlorine residues and risk of breast cancer [letter; comment] [see comments]. J Natl Cancer Inst 85:1696-7, 1993

92. Teschke K, Kelly SJ, Wiens M, et al: Concentrations of organochlorine pesticides in the adipose tissue of British Columbia residents. Can J Public Health 84:192-6, 1993
93. Mussalo-Rauhamaa H, Pantzar P: Selenium and DDE in breast fat of breast cancer patients: their relationship to hormone receptors in breast tissue [letter]. J Natl Cancer Inst 85:1964-5, 1993

4

STROMAL-EPITHELIAL CELL INTERACTIONS IN BREAST CANCER

David Heber, Judith Ashley, and Dilprit Bagga

Division of Clinical Nutrition
UCLA School of Medicine
1000 Veteran Avenue
Los Angeles, California 90095

1. INTRODUCTION

The vast majority of breast cancers originate from breast ductal epithelial cells and are either infiltrating ductal or intraductal breast cancers (1). These cells exist in a local microenvironment in which both stromal cell products and blood-borne factors may influence both transformation and tumor growth. Breast ductal epithelial cells and stromal cells have a number of things in common. Both cell types produce protein growth factors and active steroids. At the same time both cell types respond to stimulation by peptides and steroids. This chapter will explore the hypothesis that these two cell types interact in both paracrine and endocrine fashion to affect breast epithelial growth, tumor development, and breast tumor metastasis.

While all forms of cancer are genetic in that they result from genetic alterations which affect normal growth, differentiation, and apoptosis, only a minority of women have a family history of breast cancer. Since the lifetime risk for developing breast cancer is one in eight for a woman (2), it is not likely to be a purely genetic disorder. Rather breast cancer, in common with most chronic diseases, most commonly results from a genetic-environmental interaction.International comparisons of age-adjusted breast cancer incidence show marked differences with a much lower rate in Asia and developing countries than in North America, Europe, and the developed countries of the world (3,4). Substantial changes in incidence occur when populations move from low incidence to high incidence countries demonstrating that these differences are not simply genetic (5,6). The recent increase in breast cancer incidence rates in Japan suggest that environmental, dietary, and lifestyle changes within a country may also be important determinants of risk (7). Some within-country case-control studies of the relationship of dietary fat to breast cancer risk have not confirmed the findings of international correlation studies and migration studies (8). An improved understanding of events in the microenvironment of the breast ductal cell may provide clues to the growth and development of clinical breast cancer. The optimal approach to breast cancer prevention and

Dietary Fats, Lipids, Hormones, and Tumorigenesis, edited by Heber and Kritchevsky
Plenum Press, New York, 1996

control is the modification of environmental risk factors including diet, but additional information is needed on how external environmental changes are translated into factors modulating tumor development and growth. The study of the breast ductal cell microenvironment may provide such information.

2. FACTORS AFFECTING BREAST DUCTAL EPITHELIAL CELLS PROLIFERATION

The proliferation of normal breast ductal epithelium and terminal end bud growth and differentiation in the lobules of breast is the result of complex interactions between a number of hormones and growth factors including the sex steroids, insulin and various pituitary factors. Ironically, prolonged exposure to same hormones and growth factors has been implicated in the genesis and progression of breast cancer which arises from the epithelial cells of terminal ductal lobular unit. Hormones and growth factors affecting breast epithelial cell growth is shown in Table 1.

Proliferation of breast epithelial cells is predominantly controlled by the ovarian sex-steroid hormones particularly estrogens and progesterone. Cyclical exposure to these hormones in premenopausal years results in higher epithelial cell proliferation than in postmenopausal years. This augmented breast epithelial cell proliferation increases the risk of carcinogenesis by accelerating the occurrence of somatic cell genetic errors. During the postmenopausal years when there is a constant low estradiol production and very low progesterone production, there is little cell proliferation and breast cancer risk is low. Further, studies of cell kinetics in the breasts of normal premenopausal women demonstrate increased rate of cell proliferation in the mid-luteal phase of the menstrual cycle which is consistent with a proliferative stimulus from both estrogens and progesterone (9). Similarly, labeling index studies of normal breast tissues in women taking combined oral contraceptives show breast cell proliferation activity to be lowest during week 1 when contraceptives pills are not taken. However,

Table 1. Factors potentially controlling breast epithelial growth

I. Hormones
Estrogen
Progesterone
Prolactin
Insulin
Androgens
Glucocorticoids
Vitamin D
II. Growth Factors
EGF homologues (HER-2/neu, erbB-2,3,4)
Fibroblast Growth Factors (9 FGF's, 4 FGFR's)
Insulin-like Growth Factors (IGF-I, IGF-II)
Transforming Growth Factor
III. Fatty Acids
Linoleic Acid
Oleic Acid
Eicosapentaenoic Acid, Docosahexaenoic Acid

during weeks 2 to 4, total breast cell proliferation rates in women on combination oral contraceptives providing both estrogen and progesterone result in rates of cell proliferation similar to those found in breast ductal epithelium obtained from normally cycling premenopausal women (10).

It is generally considered that the proliferative response of epithelial cells to steroids is initiated by the induction of genes involved in controlling cell division or progression of cells through the cell cycle; however, one hypothesis which has received considerable attention is that steroids modulate cell proliferation by interacting with and modulating the activity of growth factors, growth factor receptors and growth factor signal transduction pathways which may be involved in breast epithelial cell proliferation.In breast cancer several growth factors and their receptors are expressed by epithelial cells or surrounding stromal cells.

Growth factors may also be important mitogens *in vivo*. Tumor cells may have altered receptor or signal transduction synthesis compared to the normal cells which may increase their responsiveness to growth factors. There are several mechanisms by which this might occur. First, circulating insulin and insulin-like growth factors (IGFs) may provide a reservoir of mitogens which could stimulate the proliferation of tumor cells by an endocrine mechanism. Second, growth factors could also be synthesized by surrounding stromal cells and increase tumor cell proliferation by a paracrine mechanism. Lastly, tumor cells may acquire the ability to synthesize growth factors which could then act as autocrine growth factors.

Epidermal growth factor (EGF) as well as transforming growth factor-alpha (TGF-alpha), both of which can activate EGF receptors (EGF-R) are probably produced locally in many tissues as local growth factors rather than systemic hormones. EGF and TGF-alpha can stimulate the growth of normal mammary epithelium and breast cancer cells *in vitro* (11).

Transforming growth factor-beta (TGF-beta) inhibits growth of breast cancer cells *in vitro* and high levels of TGF-beta expression have been shown to be positively associated with increased disease free survival (12). IGF-I and IGF-II are also potent mitogens for breast cancer cells (13). The growth effects of both are mediated predominantly via the IGF-I receptor. Although human breast cancer cell lines have been shown to secrete IGF-I (14); *in situ* hybridization demonstrated the presence of IGF-I RNA in stromal but not in normal or malignant breast epithelial cells (15). This suggests that IGF-I plays a role in breast cancer as a paracrine growth factor. IGF-II also stimulates cell proliferation by acting through the type I IGF-R and like IGF-I has been shown to be mitogenic for breast cancer cells (16,17).

Since diet can affect the lipid composition of tissues including breast stromal adipocytes, it is possible that differences in lipid composition of breast tissues may affect breast tumorigenesis. Studies of carcinogen-induced tumors in rodents, of human breast cancer growth in nude mice, and of human breast cancer cells in culture demonstrate that specific fatty acids such as linoleic acid promote tumor growth and metastasis, while other fatty acids such as oleic acid are either neutral or inhibitory (18,19).

3. BREAST ADIPOCYTES AND DUCTAL EPITHELIAL CELLS

Normal breast epithelial cell proliferation and differentiation is controlled by multiple growth factors and hormones. These include hormones such as estradiol and progesterone, prolactin, glucocorticoids and insulin; growth factors such as epidermal growth factor (EGF), insulin like growth factor (IGF) and fibroblast growth factor (FGF). Local growth regulatory peptides are elaborated by mammary stromal cells, particularly by fibroblasts, preadipocytes and adipocytes. These paracrine factors promote the growth, morphological development and differentiation of mammary epithelium (20,21). A potential role for adipocytes in growth stimulation of mammary epithelium is supported by the observation that transplanted mammary epithelium will grow only in subcutaneous sites in mice where adipose tissue is present (22).

The majority of breast cancers develop after the menopause in a milieu in which breast stromal cells have differentiated from stromal fibroblasts to adipocytes. This process of adipocyte differentiation proceeds in women between the ages of 35 and 55 coinciding with the decades during which benign breast diseases such as breast cysts and adenomas are a clinical problem.

A great deal is known about adipocyte differentiation largely derived from the obesity literature. Once an adipoblast is committed to differentiate to a preadipocyte, it begins to synthesize lipoprotein lipase (LPL) and a specific nuclear trans-acting factor (FAAR). The preadipocyte can transport fatty acids. The preadipocyte is then differentiated terminally to a mature adipocyte recognized by its accumulation of intracellular lipid droplets.

The mature adipocyte synthesizes a variety of products including: the GLUT 4 glucose transport protein, the adipsin serine protease, hormone sensitive lipase, angiotensinogen, tumor necrosis factor-alpha, fatty acid binding protein, acetyl coenzyme A synthase, and glucose phosphate dehydrogenase (23).

Adipocyte growth can be modulated by a number of factors. Insulin can affect glucose transport, lipolysis and LPL production. Glucocorticoids can affect fat distribution and rates of lipolysis. IGF-I and IGF-II can affect growth through their actions on the insulin receptor (24). Finally, fatty acids can affect adipocyte growth and differentiation. Fatty acids can act as nutrients, but can also affect membrane composition and function. They can affect prostaglandin production, ion channels, and enzymatic function. They can affect neural transmission across synapses, and studies have demonstrated that fatty acids can affect gene expression. In fact, fatty acids are in the family of amphipathic carboxylates which includes 9-cis-retinoic acid. Palmitate has been shown to affect gene expression in adipocytes (25).

Adipocytes are a major source of free fatty acids. Furthermore both adipocytes and muscle cells have been long known as the primary source of LPL. Mammalian adipose tissue is also a major source of active CETP and apolipoprotein E (apo E). Adipose tissue is known to contain the major cholesterol pool of the body and the locally synthesized apo E and CETP helps in removal of adipocyte cholesterol. The capacity of adipose tissue as a secretory organ extends to sex steroid synthesis. The product of P-450 aromatase mRNA which is required for the synthesis of estrone and estradiol is expressed in adipose tissue (26). Breast adipocytes in common with adipocytes from subcutaneous fat can convert androstenedione to estrone and secrete a number of peptides as reviewed above. The possibility that breast fat may influence breast cancer development has been raised previously. This peripheral steroid formation may thus be quantitatively important owing to the differences in fat tissue in lean and obese patients. Since dietary interventions which reduce fat and calorie intake can also lead to reductions in body fat and breast fat. It is possible that increases in breast fat observed in obese postmenopausal women promote breast cancer development. Increased mammographic density has also been identified as a strong risk factor for breast cancer development in premenopausal women (27). The mammographic density is determined by the ratio of radiographically dense stromal-vascular-epithelial components to the more lucent adipose tissue of the breast. Both estrogen and peptide growth factors are produced by the stromal cells of the breast. After age 35 these cells begin to differentiate into fat cells, and after menopause the breast is fatty containing primarily atrophic breast ductal epithelial tissue.

4. OBESITY AND BREAST CANCER RISK

Up to 45% of women over the age of 45 are classified as obese, and it is common for body fat to redistribute to the upper body after menopause (28). Since women with upper body obesity may have an increased risk for breast cancer (29), there is a possibility that breast fat may play a role in breast cancer promotion. These women often have increased breast fat and that fat tissue may actively secrete substances locally or systemically which affect tumor development and growth.

Increased estrogen production in women with excess fat compared to lean women (30-32) is due to increased peripheral conversion of androstenedione to estrone by fat tissue (33-35). Obesity has been associated repeatedly with more advanced breast cancer at the time of diagnosis, higher rates of recurrence, and shorter survival times even after controlling for tumor size and stage of disease at diagnosis (36-41). This has been explained in the past as resulting from delayed diagnosis in obese women rather than enhanced tumor promotion resulting from increased estrogen production. The effects of obesity on the invasiveness of breast cancers at the time of diagnosis are different among patients with estrogen receptor-positive tumors compared to those with estrogen receptor-negative tumors (42). If the effects of obesity on breast tumor nodal invasion and, therefore, prognosis were due simply to delayed diagnosis then similar associations should have been seen for both the receptor-positive and negative patients. The fact that such an effect of obesity was observed only in the receptor positive patients strongly suggests that a hormonal mechanism accounts for the effects of obesity on breast tumor promotion.

Studies of changes in circulating hormone levels in premenopausal women with marked changes in body weight have been complicated by the occurrence of inadequate luteal phases observed in markedly obese women as co-existing hyperandrogenism including the insulin resistance-hirsutism syndromes (viz. polycystic ovarian syndrome). Older studies have used weight and height only, but there are no careful studies of body composition changes in obese premenopausal women to document the effects on circulating estrogens.

There are two sources of estrogen in premenopausal women, ovarian secretion and the conversion of adrenal androgens to estrogens by body fat. This conversion is carried out by the stromal cells of fat tissue through the action of aromatase, a cytochrome P-450 oxidase. This enzyme can be induced *in vitro* by dexamethasone and insulin (43). Since aromatization of androgens is the only source of circulating estrogens in postmenopausal women, modulation of aromatase activity may be an important means of changing estrogen levels. In our studies (44) of postmenopausal women eating a high fiber, low fat diet, women lost small amounts of weight over a three week period but did not decrease their fat mass significantly. Nonetheless, we observed a 40% decrease in circulating estradiol levels suggesting that a modulation of aromatase activity can also be observed prior to a marked decrease in fat mass.

Breast cancer has also been associated with increased percent free estradiol and reduced SHBG levels (45-49). Increased concentrations of free fatty acids are associated with increased percent free estradiol (50) most likely by displacing estradiol from SHBG. Estradiol is bound to both Sex Hormone Binding Globulin (SHBG) and albumin. It has been proposed that only the unbound fraction is available for transport into tissues. Pardridge and Mietus demonstrated that estradiol bound to SHBG did not cross endothelia of peripheral capillaries (51). Other investigators maintain that there is little effect of SHBG on bioavailable estradiol since its affinity for estradiol is much lower than for testosterone. Only about 50% of estradiol is bound to SHBG. Nonetheless, some investigators have found breast cancer to be associated with increased percent free estradiol and reduced SHBG levels (45-49), and increased concentrations of free fatty acids correlate with increased percent free estradiol (50) most likely by displacing estradiol from SHBG. Abdominal obesity is also associated with decreased SHBG in postmenopausal women, and the SHBG levels are increased following weight reduction.Therefore, upper body obesity may put post-menopausal women at increased risk via increases in serum estradiol. Conversely, these women may have a greater decrease in breast cancer risk with dietary modification as suggested by Schapira (52). In both men and postmenopausal women, weight loss results in decreased estrogens levels over a few weeks prior to major weight loss.

Several studies have demonstrated that breast cancer patients are more likely to have abdominal obesity. Increased waist-to-hip ratio has been associated strongly with breast cancer in case-control studies (53,54). Increased abdominal obesity is associated with increased androgen levels, increased levels of free fatty acids, increased bioavailable estrogen, and decreased levels of Sex Hormone Binding Globulin (55). Abdominal adipocytes have a higher basal rate of free fatty acid release compared to thigh adipocytes (55).The relative risks of breast cancer are increased significantly for any given level of obesity by considering fat distribution, and a reduction in intra-abdominal fat will theoretically reduce risk significantly (56). Increased waist-to-hip ratio (and increased abdominal fat) have been correlated with the variability in androgen levels in women (57). The increased androgenicity and bioavailable estrogen levels observed with this fat distribution mirror the findings of certain case-control studies suggesting that abdominal obesity may be another marker of susceptibility to the promotional effects of a high fat, high calorie diet.

While obesity on a weight for height basis has not been associated with an increased risk of premenopausal breast cancer, adult weight gain increases the risk of postmenopausal breast cancer and is a predictor of breast cancer risk independent of body weight (58). In our view, the physiological effects of weight gain will be evident in premenopausal as well as in postmenopausal women.

5. DIETARY INTAKE AND ESTROGEN LEVELS

A number of aspects of dietary intake and body composition affect endogenous estrogen levels. High fat diets are associated with obesity in genetically susceptible women, and obesity has been identified in epidemiological studies as a significant risk factor for breast cancer (59). Women eating a 40% fat diet ad libitum ingest 600 Cal/day more than women eating a 20% fat diet (60). Dietary fat reduction results in a pattern of changes in diet including changes in energy intake and fiber. Usual diets in postmenopausal women in the U.S. are not only higher in fat than diets in countries with a low incidence of breast cancer (e.g. China) but are also lower in fiber and micronutrients. Diets rich in fruits and vegetables should lead to increased circulating levels of carotenoids, vitamin C and E. However, fiber can interfere with absorption of micronutrients in parallel to its effects on steroid metabolism.

Epidemiological, experimental and clinical data have been published supporting the concept that increasing dietary fiber may help prevent breast cancer (61-63). Fiber could exert its protective effects by changing the metabolism of estrogens resulting in less circulating bioavailable estrogen in two ways. First, the fibers could directly bind estrogens taking them out of the circulation via the feces. Second, it has been proposed that certain bacterial flora populations may change as the result of dietary modification. Those bacteria normally associated with a high fat,low fiber diet which deconjugate estrogen conjugates may be reduced in number resulting in reduced unconjugated estrogens for recirculation via the enterohepatic pool. Studies of omnivores compared to vegetarians in the Boston area demonstrated that the vegetarians had lower levels of estrogen in serum and secreted more estrogen in their feces (63). There was an inverse relationship between urinary and fecal estrogen concentrations.

6. EVIDENCE OF STROMAL-EPITHELIAL INTERACTION

The stroma is in close association with the secretory epithelium surrounding the terminal ductal lobular unit. The intralobular stroma is loose, cellular and highly vascular, whereas the extralobular stroma is dense fibrous tissue in the mammary gland (64). The cell fraction associated with the mammary stroma are fibroblasts, preadipocytes and adipocytes. Paracrine influences whereby surrounding stromal tissues secrete factors that interact with the mammary epithelial cells may be of more importance that previously presumed.

PDGF is a potent mitogen for fibrobalstic cells (65) but not for the breast tumor cells (66). The mitogenic effects of tumor derived PDGF may be mediated through stimulation of IGF production , since PDGF stimulates IGF-I production in human fibroblast cells (65). The IGF-I in turn is a potent mitogen for both fibroblast cells and breast cancer cells (13). Thus, the paracrine loops between the stromal and breast tumor cells may have a reciprocal proliferative effect.

Simpson et al (67) have demonstrated that breast stromal cells' aromatization of androstenedione decreases with increasing distance from a breast tumor suggesting that a paracrine signal from the tumor cells is stimulating aromatization by the nearby stromal cells.

Our group has studied breast stromal adipocytes obtained from breast cancer patients (35). In these studies, breast tissue samples were transported in chilled culture medium on ice from the operating room to the laboratory. The fibroblast stromal fraction was separated after collagenase digestion and centrifugation. The cells were cultured in serum-free medium for 24 hours. Then the cells were cultured for an additional 72 hours in the presence of dexamethasone, dibutyryl cyclic AMP , and phorbol myristate acetate. Cells obtained from 4 patients demonstrated marked stimulation following the above treatment (see table 2).

Further studies of the control of aromatase in these cells is shown in figures 1A and 1B below where increasing concentrations of phorbol myristate acetate stimulated significant dose-related increases in aromatization even at maximally stimulating doses of dibutyryl cyclic AMP.

Table 2. Human Breast Aromatase
(raw activity in counts per minute)

Patient No.	Baseline	Stimulated
1	155 ± 7	1075 ± 162
2	238 ± 15	4410 ± 1029
3	150 ± 33	6120 ± 3271
4	28 ± 12	1247 ± 359

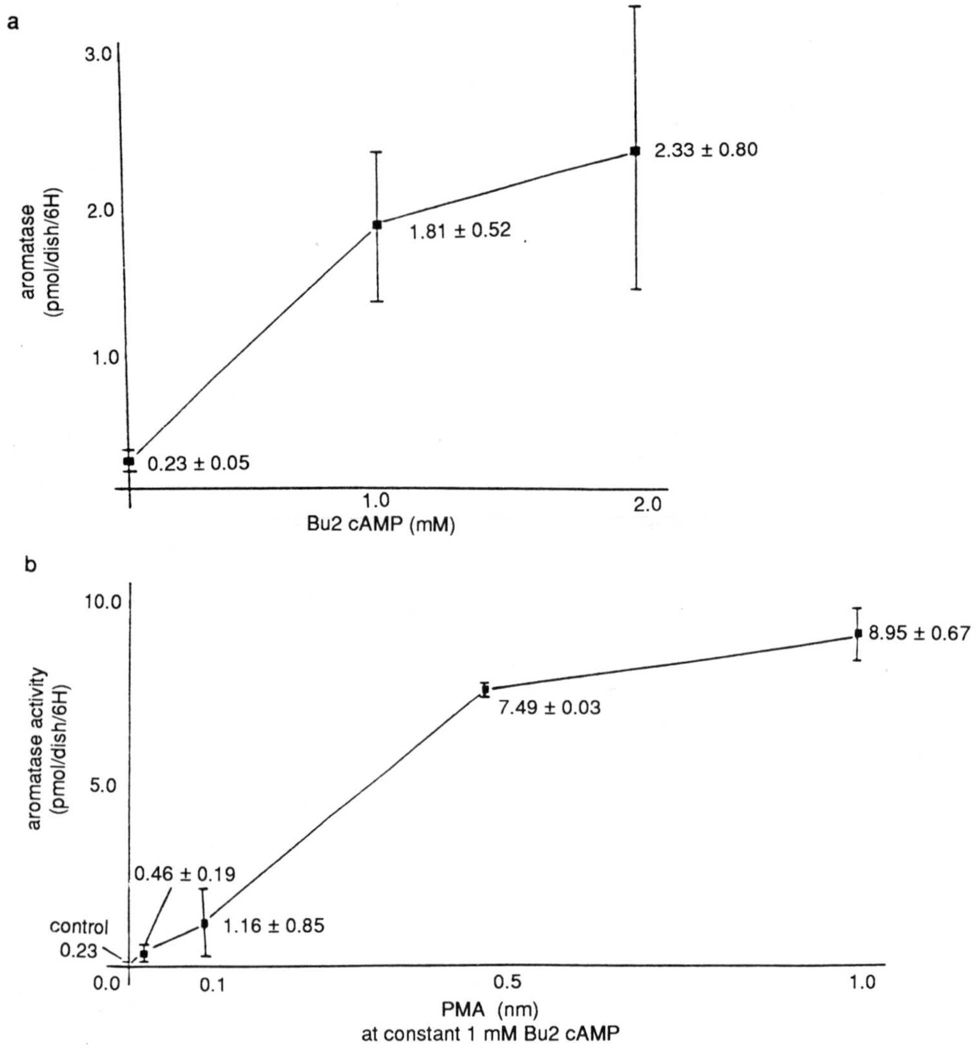

Figure 1. (a) Increasing concentrations of dibutyryl cyclic AMP were incubated for 72 hours with cultured breast adipose stromal cells from a patient undergoing mastectomy. Points shown are based on duplicate determinations. Units are shown on the abscissa and ordinate for aromatase activity in pmol/dish/6 hours and for dibutyryl cyclic AMP in mM concentration used in the co-incubation study. (b) Increasing concentrations of phorbol myristate acetate were incubated for 72 hours with cultured breast adipose stromal cells from a patient undergoing mastectomy. A constant maximally stimulating concentration of dibutyryl cyclic AMP was co- incubated with the cells for 72 hours. PMA concentration is shown on the abscissa in nanomoles per liter. Since a maximally stimulating dose of dibutyryl cyclic AMP was used throughout these data suggest that a synergistic stimulatory effect via a different mechanism results form PMA treatment of these cells.

Spicer et al (68) have shown that the stromal-vascular-epithelial complex seen by mammography will respond to hormonal modulation. Women with significant fibrocystic changes demonstrate decreased cellular tissue with pharmacologic ovariectomy using a gonadotropin-releasing hormone agonist. Johnston et al (69) have shown that conditioned medium from a 3T3L1 mouse preadipocyte cell line can inhibit growth and thymidine incorporation of cultured human MCF-7 breast cancer cells. Conditioned media from the parent 3T3 fibroblast cell line led to stimulation of breast cancer cell growth and thymidine incorporation. These studies suggest that the adipocytes in the breast stromal environment can modulate tumor growth.

7. FUTURE DIRECTIONS FOR RESEARCH

There are a number of directions for future research which may prove fruitful. First, there is the study of the modulation of breast cancer cell differentiation by adipocytes. These studies need to be performed with both estrogen-sensitive and estrogen-independent cell lines. In addition, studies with immortalized normal epithelial cells or primary breast cell culture may help to elucidate the differences between normal and tumor tissue in regard to sensistivity to stimulation by adipocytes. Once this phenomenon has been demonstrated then the reverse studies of modulation of adipocyte differentiation by breast cancer cells should be examined.

Finally, the effects of various modulating agents on the combined breast ductal cell-stromal cell unit should be examined. The ultimate goal of these studies must be to identify and characterize all of the factors involved in stromal-epithelial cell interactions. This information could provide significant insights into dietary and hormonal strategies for breast cancer prevention and treatment.

8. REFERENCES

1. American Cancer Society. Cancer Facts and Figures. Atlanta, GA. American Cancer Society, 1993.
2. Feuer EJ, Wun LM, Boring CC, Flanders WD, Timmel MJ, Tong T. The lifetime risk of developing breast cancer. J Natl Cancer Inst 1993; 85:892-297.
3. Henderson BE, Bernstein L. The international variation in breast cancer rates: an epidemiological assessment. Breast Cancer Research and Treatment 1991; 18 Suppl 1:S11-17.
4. Miller AB, Bulbrook RD. UICC multidisciplinary project on breast cancer: the epidemiology, aetiology and prevention of breast cancer. Int J Cancer 1986; 37:173-177.
5. Shimizu H, Ross RK, Bernstein L, Yatani R, Henderson BE, Mack TM. Cancer of the prostate and breast among Japanese and white immigrants in Los Angeles County. Br J Cancer 1991; 63:963-966.
6. Tsugane S, de Souza JMP, Costa ML Jr, Mirra JP, Gottlieb SLD, Laurenti R, Watanabe S. Cancer incidence rates among Japanese immigrants to Sao Paulo, Brazil, 1969-1978. Cancer Causes and Control 1990; 1:189-193.
7. Kodama M, Kodama T, Miura S, Yoshida M. Nutrition and breast cancer risk in Japan. Anticancer Research 1991; 11:745-754.
8. Willett WC, Hunter DJ, Stampfer MJ et al. Dietary fat and fiber in relation to risk of breast cancer: an 8 year follow-up. JAMA 1992; 268: 2037-2044.
9. Pike MC, Spicer DV, Dahmoush L, and Press MF. Estrogens, Progestogens, normal breast cell proliferation and breast cancer risk. Epidemio Rev 1993; 15: 17-35.
10. Anderson TJ, Batterbys S, King RJB et al. Oral contraceptive use influences resting breast proliferation. Human Pathol 1989; 20: 1139-1144.
11. Klokm JGM, Berns PMJJ, van Putten WLJ, Bontenbal M et al. Critical review of growth factors as clinical tools in primary and metastatic breast cancer. Recent Results Canc Res 1993; 127:77-88.
12. Coombes RC, Barrett–Lee P, Luqmani Y. Growth factor expression in breast tissue. J Steroid Biochem Molec Biol 1990; 37: 833-837.
13. Westley BR and May FEB. Role of Insulin-like growth factor in steroid modulated proliferation. J Steroid Biochem Molec Biol 1994; 51: 1-9.
14. Freed KA, Herrington AC. Insulin-like growth factor-I and its autocrine role in growth of MCF-7 human breast cancer cells in culture. J Molec Endoc 1989; 3: 183-190.
15. Yee D, Paik S, Lebovic GS, Lippman ME et al. Analysis of insulin-like growth factor I gene expression in malignancy: evidence for a paracrine role in human breast cancer. Molec Endocr 1989; 3: 509-517.

16. Stewart AJ, Johnson MD, May FEB et al. Role of insulin-like growth factor and the type I insulin-like growth factor receptor in the estrogen stimulated proliferation of human breast cancer cells. J Biol Chem 1990; 34: 21172-21178.
17. Yee D, Cullen KJ, Paik S et al. Insulin-like growth factor -II mRNA in human breast cancer. Cancer Res 1988; 48: 6691-6696.
18. Carroll, KK, Khor HT Effect of level and type of dietary fat on incidence of mammary tumors induced in female Sprague-Dawley rats by 7,12 dimethylbenzanthracene. Lipids 1971; 6:415-419.
19. Rose DP, Connolly JM. Effects of fatty acids and inhibitors of eicosanoid synthesis on the growth of human breast cancer cell lines in culture. Cancer Res 1990; 50:7139-7144.
20. Lyons WR, Li CH, Johnson RE. The hormonal control of mammary growth and lactation. Recent Prog Hormone Res 1958; 14: 219-254.
21. Deeks S, Richards J, Nandi S. Maintainance of normal rat mammary epithelial cells by insulin and insulin-like growth factor I. Exp Cell Res 1988; 174: 448-460.
22. Sheffield LG and Welsh CW. Transplantation of human breast epithelia to mammary-gland-free fat pads of athymic nude mice: influence of mammographic hormones on growth of breast epithelia. Int J Cancer 1988; 41: 713-719.
23. Ailhaud G, Grimaldi P and Negrel R. Cellular and molecular aspects of adipose tissue development. Ann Rev Nutr 1992; 12: 207-33.
24. Cornelius P, MacDougald OA, Lane MD. Regulation of adipocyte development. Ann Rev Nutr 1994; 14: 99-129.
25. Amri EZ, Ailhaud G, Grimaldi PA. Fatty acids as signal transducing molecules: involvement in differentiation of preadipose to adipose cells. J Lipid Res 1994; 35: 930-937.
26. Labrie F, Simard J, Luu-The V, Trudel C, Martel C et al. Expression of 3beta-hydroxysteroid dehydrogenase/delta4-delta5 isomerase (3 beta-HSD) and 17 beta-hydroxysteroid dehydrogenase (17 beta-HSD) in adipose tissue. Int J Obes 1991; 15: 91-99.
27. Boyd NF, O'Sullivan B, Campbell JE et al. Mammographic signs as risk factors for breast cancer. Br J Cancer 1982; 45: 185-193.
28. Bouchard C, Bray GA, Hubbard vs. Basic and clinical aspects of regional body fat distribution. Am J Clin Nutr 1990; 52:946-950.
29. Schapira DV, Clark RA, Wolff PA, Jarrett AR, Kumar NB, Aziz NM. Visceral obesity and breast cancer risk. Cancer 1994; 74:632-639.
30. Bernstein L, Henderson BE, Hanisch R, Sullivan-Halley, Ross RK. Physical exercise and reduced risk of breast cancer in young women. J Natl Cancer Inst 1994; 86: 1403-1408.
31. Goldin JM, Siiteri PK, MacDonald PC. Source of estrogen production in postmenopausal women. J Clin Endocrinol Metab 1973; 36:207-210.
32. Siiteri PK, MacDonald PC. Role of extraglandular estrogen in human endocrinology. In: Greep RO, Astwood EB, eds. Handbook of Physiology Sec.7, vol.2, part 1 Wash. D.C. Am Physiol Soc 1973; 15: 629.
33. Schindler AE, Ebert A, Friedrich E. Conversion of androstenedione to estrone by human fat tissue. J Clin Endo Metab 1972; 35:627-630.
34. Simpson ER, Ackerman GE, Smith ME. Estrogen formation in stromal cells of adipose tissue of women: induction by glucocorticoids. Proc Nat Acad Sci 1981; 73:5690-5694.
35. Heber D, Koziol BJ, Shoshani A, Su SY, Colburn J, Rosser RJ, Waisman J, and Silverstein M. Regulation of breast tissue estrogen production by tumor-associated rat and human stromal fibroblasts: cell culture studies. Abst. 69th Annual Mtg. of the Endocrine Society. The Endocrine Society, Rockville, MD 1987.
36. Zumoff B. Relationship of obesity to blood estrogens. Cancer Res.(suppl) 1982; 42:3289-3294.
37. Kalish L. Relationship of body size with breast cancer. J Clin Oncol 1984; 2:287-293.
38. Abe R, Kumagai N, Kimura M. Biological characteristics of breast cancer in obesity. Tohoku J Exp Med 1976; 120:351-9.
39. Donegan WL, Hartz AJ, Rimm AA. The association of body weight with recurrent cancer of the breast. Cancer 1978; 41:1590-1594.
40. Howson CP, Kinne D, Wynder EL. Body weight, serum cholesterol and stage of primary breast cancer. Cancer 1986; 58:2372-2381.
41. De Waard F Epidemiology of breast cancer: a review. Eur J Clin Oncol 1983; 19:1671-1676.
42. Verreault R, Brisson J, Deschenes L, Naud F. Body weight and prognostic indicators in breast cancer. Am J Epidem 1989; 129: 260-268.
43. Ackerman GE, Smith ME, Mendelson CR, MacDonald PC, Simpson ER. Aromatization of Androstenedione by Human Adipose Stromal Cells in Monolayer Culture. J Clin Endo Metab 1981; 53: 412-417.

44. Heber,D, Ashley JM, McCarthy, WJ, Solares, ME, Leaf,DA, Chang, L-J C, and Elashoff RM. Assessment of adherence to a low fat diet for breast cancer prevention. Prev Med 1992; 21: 218-227.
45. Siiteri PK, Hammond GL, Nisker JA Increased availability of serum estrogens in breast cancer: a new hypothesis. In: Pike MC, Siiteri PK, Welsch CW, eds. Hormones and Breast Cancer. Banbury report no. 8 Cold Spring Harbor, NY. Cold Spring Harbor Laboratory, 1981:87-106.
46. Moore JW, Clark GMG, Bulbrook RD, et al. Serum concentrations of total and non-protein bound estradiol in patients with breast cancer and in normal controls. Int J Cancer 1986; 38:625-630.
47. Ota DM, Jones LA, Jackson GL, et al. Obesity, non-protein bound estradiol in the sera of breast cancer patients. Cancer 1986; 57:558-562.
48. Bruning PF, Bonfrer JMG, Hart AM. Non-protein bound oestradiol, sex hormone binding globulin, breast cancer, and breast cancer risk. Br J Cancer 1985; 51: 479-484.
49. Ernster VL, Wrensch MR, Petrakis NL, King EB, Miike R, Murai A, Goodson, WH, and Siiteri, PK. Benign and malignant breast disease: initial study results of serum and breast fluid analyses of endogenous estrogens. J Natl Cancer Inst 1987; 79:949-960.
50. Bruning PF, Bonfrer JMG. Free fatty acid concentrations correlated with the available fraction of estradiol in human plasma. Cancer Res 1986; 46:2606-2609.
51. Pardridge WM, Mietus LJ. Transport of steroid hormones through the rat blood-brain barrier. Primary role of albumin-bound hormone. J. Clin. Invest. 1979; 64:145-154.
52. Schapira DV, Kumar NB, Lyman GH Abdominal obesity and breast cancer risk. Ann Int Med 1990; 112:182-186.
53. Ballard-Barbash R, Schatzkin A, Carter CL, et al. Body fat distribution and breast cancer in the Framingham study. J Natl Cancer Inst 1990; 82:286-290.
54. Folsom AR, Kaye SA, Prineas RJ et al. Increased incidence of carcinoma of the breast associated with abdominal adiposity in postmenopausal women. Am J Epidemiol 121: 794-803.
55. Kissebah AH, Evans DJ, Peiris A, Wilson CR. Endocrine characteristics in regional obesities: role of sex steroids. In Vague J, Bjorntorp P, Guy-Grand B, eds. Metabolic complications of human obesities. Elsevier Science Publishers, Amsterdam, 1985; 115-130.
56. Schapira DV, Kumar NB, Lyman GH. Estimate of breast cancer risk reduction with weight loss. Cancer 1991; 67:2622-2625.
57. Kaye SA, Folsom AR, Soler JT, Prineas RJ, Potter JD. Associations of body mass and fat distribution with sex hormone concentrations in postmenopausal women. Int. J. Epidemiol. 1991; 20:151-156.
58. Harris RE, Namboodiri KK, Wynder E 1992 Breast cancer risk: effects of estrogen replacement therapy and body mass. J Natl Cancer Inst 1992; 84:1575-1582.
59. LeMarchand L, Kolonel LN, Earl ME, Mi M-P. Body size at different periods of life and breast cancer risk. Am J Epidemiol 1988;128:137-152.
60. Kendall A, Levitsky D, Strupp BJ, Lissner L. Weight loss on a low fat diet: consequence of the imprecision of the control of food intake in humans. Am J Clin Nutr 1991; 53:1124-1129.
61. Hirayama T: A large scale cohort study on cancer risks by diet - with special reference to the risk reducing effects of green-yellow vegetable consumption. In Diet,Nutrition and Cancer, Y Hayashi et al eds. , Japan Sci Soc Press, Tokyo/VNU Sci Press, Utrecht 1986, pp. 41-53.
62. Adlercreutz H, Gorbach SL, Woods M et al. The relationship between estrogen levels and diets of caucasian American and oriental immigrant women. Am J Clin Nutr 1986; 44:945-953.
63. Goldin BR, Adlercreutz H, Gorbach SL, Warram JH et al. Estrogen excretion patterns and plasma levels in vegetarian and omnivorous women. New Engl J Med 1982; 307:1542-1547.
64. Azzopardi JG. Problems in breast pathology. London: Saunders, 1979; 8-15.
65. Clemmons DR, Shaw DS. Variables controlling somatomedin production by cultured human fibroblasts. J Cell Physiol 1983; 115: 137-140.
66. Bronzert DA, Pantazis P, Antoniades HN et al. Synthesis and secretion of platelet-derived growth factor by human breat cancer cell lines. Proc Natl Acad Sci 1987; 84: 5763-5769.
67. Balun SE, Mahendroo MS, Simpson ER. Aromatase gene expression in adipose tissue: relationship to breast cancer. J Steroid Biochem Molec Bio 1994; 49: 319-326.
68. Spicer DV, Ursin G, Parisky YR, Pearce JG, Shoupe D et al. Changes in mammographic densities induced by a hormonal contraceptive designed to reduce breast cancer risk. J Natl Cancer Inst 1994; 86: 431-436.
69. Johnston PG, Rondinone CM, Voeller D, Allegra CJ. Identification of a protein factor secreted by 3T3-L1 preadipocytes inhibitory for the human MCF-7 breast cancer cell line. Cancer Res 1992; 52: 6860-6865.

5

20-EPI-VITAMIN D_3 ANALOGS

Potent Modulators of Proliferation and Differentiation of Breast Cancer Cell Lines *in Vitro*

E. Elstner,[1] D. Heber,[2] and H. P. Koeffler[1]

[1] Div. of Hematology-Oncology, Cedars–Sinai Medical Center
UCLA School of Medicine
8700 Beverly Blvd., D5033
Los Angeles, California 90048
[2] Division of Clinical Nutrition
UCLA School of Medicine
1000 Veteran Avenue
Los Angeles, California 90095

1. ABSTRACT

Breast cancer is a devastating disease. New approach in therapy are needed. In this chapter, we describe our efforts to identify novel vitamin D_3 analogs which may have a potent antiproliferative effect on breast cancer without causing hypercalcemia.

2. INTRODUCTION

The physiologically active 1,25 dihydroxyvitamin D_3 [1,25$(OH)_2D_3$] causes intestinal calcium absorption and bone calcium mobilization. The mechanism of action of 1,25$(OH)_2D_3$ is likely similar to that of other steroid hormones. The 1,25$(OH)_2D_3$ interacts noncovalently, but stereospecifically with intracellular receptor proteins. This steroid receptor complex associates with DNA in the nucleus (vitamin D response elements) either to initiate or inhibit RNA synthesis of selective genes[1].

Indirect and direct evidence suggest that vitamin D_3 can play a role in modulating development and growth of breast cancer. An inverse correlation exists between annual exposure to development and growth of breast cancer[2,3]. The serum levels of VD_3 are related to exposure to sunlight because vitamin D is synthesized from 7-dehydrocholesterol in the skin after exposure to sunlight. Further epidemiologic studies have suggested that a low serum level of VD_3 is a risk factor for breast tumors express high affinity intracellular 1,25$(OH)_2D_3$ receptors (VDR)[5-7]. Patients with primary carcinoma of the breast who were VDR-positive had significantly longer disease-free survival than those with receptor-nega-

Dietary Fats, Lipids, Hormones, and Tumorigenesis, edited by Heber and Kritchevsky
Plenum Press, New York, 1996

tive tumors[8]. The VDR belongs to the same steroid receptor superfamily as ER[9] and is the smallest of the steroid hormone receptor[10,11]. The properties of VDR in breast cancer cells are similar to those of receptors present in normal mammalian target organs such as the bone and intestine[12]. Unlike estrogen, which stimulates tumor growth, $1,25(OH)_2D_3$ and its analogs inhibit proliferation of breast cancer cells *in vitro*[13-15]. Likewise, $1,25(OH)_2D_3$ and related compounds decrease the development and progression of breast cancer and other carcinomas *in vivo*[16,17], inhibit metastatic spread of tumor cells[8,18-22] and promote differentiation of breast cancer cells as well as other cancer cell types irrespective of ER status of the tumor cells[8,15,23-34]. During the last ten years, we have published over 60 laboratory and clinical studies on the effect of $1,25(OH)_2D_3$ and vitamin D analogs on differentiation and proliferation of cancer cells[1,22,31,34-91]. Because the major toxicity of $1,25(OH)_2D_3$ is hypercalcemia, we and other investigators have turned our attention to development and study of new vitamin D analogs that induce differentiation and inhibit proliferation of cancer cells without causing hypercalcemia[14,16,18,22,75,80,87,88,91,92]. We have *analyzed* the effect of over *280 vitamin D analogs* on cancer cells and have found that about 30 of these compounds were more active than $1,25(OH)_2D_3$ in induction of differentiation and inhibition of proliferation of malignant cells and yet, they also had between 10- and 300-fold less ability to induce hypercalcemia (Figure 1).

Several of the vitamin D analogs, including $1,25(OH)_2$-16ene-23yne-D_3 have been singled out by us for further studies because they have marked potency and yet have 10- to 25-times less ability than $1,25(OH)_2D_3$ to cause hypercalcemia[22]. We developed several cancer models using malignant cells injected into mice. In each model system, several of the analogs were able to increase significantly the survival of the mice as compared to those who received either $1,25(OH)_2D_3$ or diluent[22]. We and others have shown that $1,25(OH)_2D_3$ can inhibit the *in vitro* proliferation of human breast cancer cell lines[14,16,18,91].

In this chapter, we report that the 20-epi-$1,25(OH)_2D_3$ compounds are extremely potent in their antiproliferative effects on breast cancer cell lines.

3. MATERIALS AND METHODS

3.1. Cell Lines

The breast cancer cell lines (MDA-MB-436, MCF-7, SK-B-BR-3, BT474, BT20, MDA-MB-231) were obtained from ATCC (Rockville, MD). The cells were cultured according to ATCC recommendations in culture flasks with vented filter caps (Costar, Cambridge, MA).

3.2. Vitamin D_3 Compounds

The vitamin D_3 compounds were dissolved in absolute ethanol at 10^{-3} M as a stock solution, which was stored at -20° C and protected from light. The $1,25(OH)_2D_3$ ($1,25D_3$) was synthesized by Hoffman-LaRoche Inc., Nutley, NJ. The KH 1060 [20-epi-22oxa-24a,26a,27a-tri-homo-$1,25(OH)_2D_3$] and MC 1288 [($1,25(OH)_2$-20epi-D_3] were synthesized in the Department of Chemical Research, Leo Pharmaceutical Products, Denmark.

3.3. Clonogenic Assay in Soft Agar

Breast cancer cells were cultured in a two-layer soft agar system for 14 days without adding any growth factors as previously described[93].

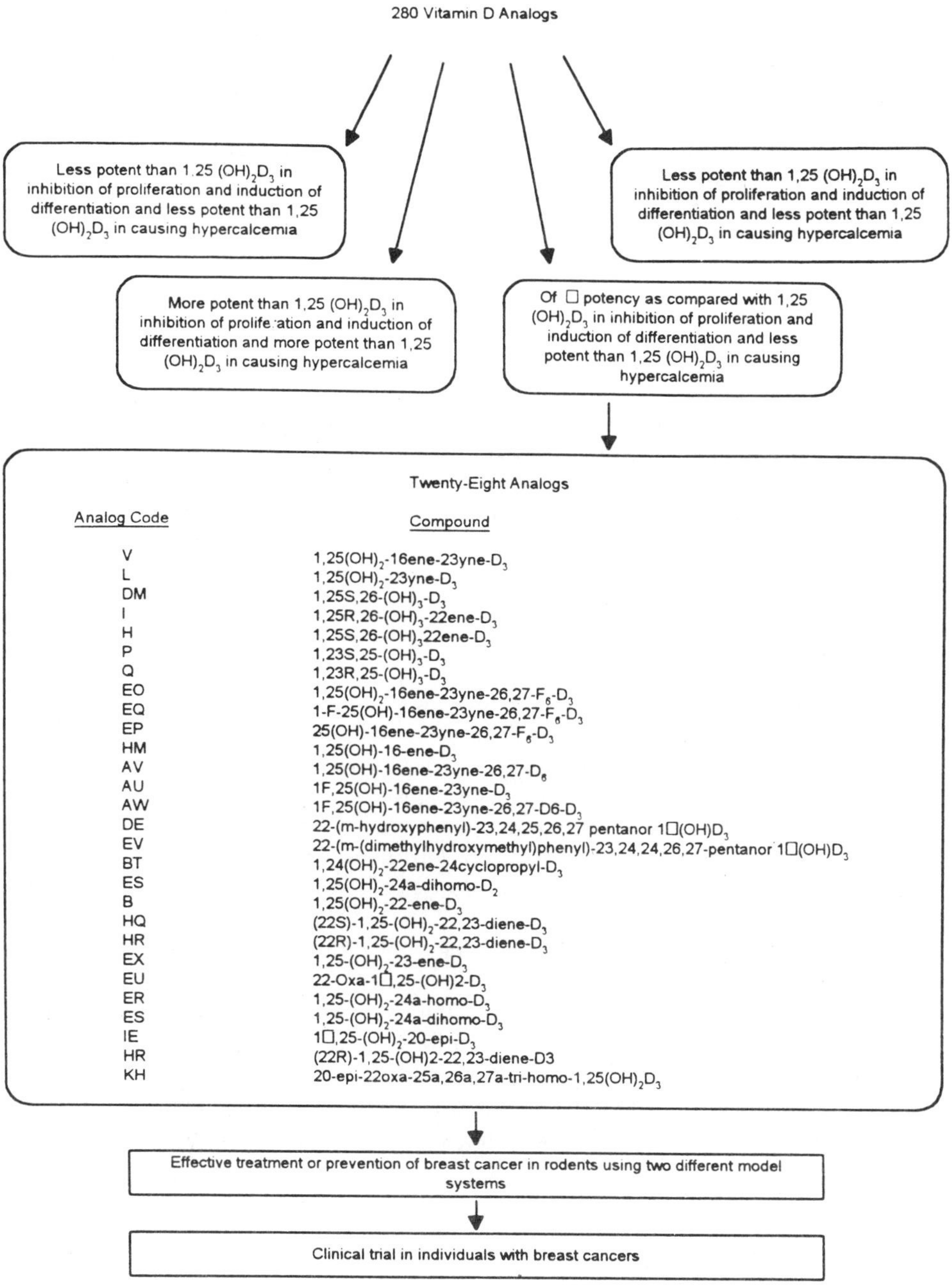

Figure 1. Evaluation of Vitamin D analogs.

Table 1. Expression of VDR, ER, p53, and bcl-2 in breast cancer cell lines

Breast cancer cell lines	VDR*	ER** staining intensity***	p53 mutational status°	bcl-2 staining intensity
MCF-7	+	+++	wt	++
BT-474	+	+	mut	+++
MDA-MB-231	+	-	mut	+
SK-BR-3	+	-	mut	-
BT-20	+	-	mut	+
MDA-MB-436	+	-	N.I.	+++

*Data not shown
** Estrogen receptor (ER) data from (43) and Dr. Grund (personal communication).
***Staining intensity: negative (-) or + weak, ++ moderate, +++ strong positivity.
°data for type of p53 from (44) and (45)
N.I.-no information; wt, wild type p53; mut, mutant p53

3.4. Induction of Differentiation in Breast Cancer Cell Lines

The breast cancer cell lines contain a small fraction of cells that exhibit morphological maturation and differentiation-markers associated with milk production; they contain large lipid droplets[94]. Breast cancer cell lines were cultured on chamber slides at a density of 2×10^5 in 3 ml medium. At 24 hours after seeding at a time when cell adhesion was complete, $1{,}25D_3$ and its analogs were added at 10^{-7} M for 3 days without medium change. For lipid visualization, we used a modified Oil Red O in propylene glycol method[95,96].

3.5. Apoptosis

Apoptosis was assessed by analyzing changes in cell morphology and DNA fragmentation after 24, 48, and 72 hours of incubation of breast cancer cell lines with KH 1060 at 10^{-6} M. Morphologically, cells undergoing apoptosis were identified by intense staining, highly condensed chromatin, fragmented nuclear chromatin, a general decrease in overall cell size, and cellular fragmentation into apoptotic bodies[97]. These changes were readily observed in cytospin preparations stained with Diff-Quick Stain Set (Baxter Healthcare Corporation, Miami, FL). Enumeration of apoptotic cells was performed by evaluation of about 300 cells by light microscopy.

The extent of DNA fragmentation was determined by a modification of the method of Sellins and Cohen[98]. Positive control for DNA fragmentation was a human T-cell leukemia cell line (CEM) treated for 8 hours with anti-CD3 mAb (OK T3) (Ortho Biotech, Raritan, NJ) (Umiel, submitted) and MCF7 cells cultured for 6 days without medium change.

To detect apoptotic cells in our samples, we also used the Oncor Apop Tag™ in Situ Apoptosis Detection Kit (Oncor, Gaithersburg, MD) directly detecting digoxigenin labeled genomic DNA by flow cytometry. The labeling targets of the Apop Tag Kit are new 3′-OH DNA ends generated by DNA fragmentation, typically localized in morphologically identifiable nuclei and apoptotic bodies.

3.6. Immunohistochemistry

Immunostaining for bcl-2 and p53 was performed on breast cancer cells which had been grown in culture with and without KH 1060.

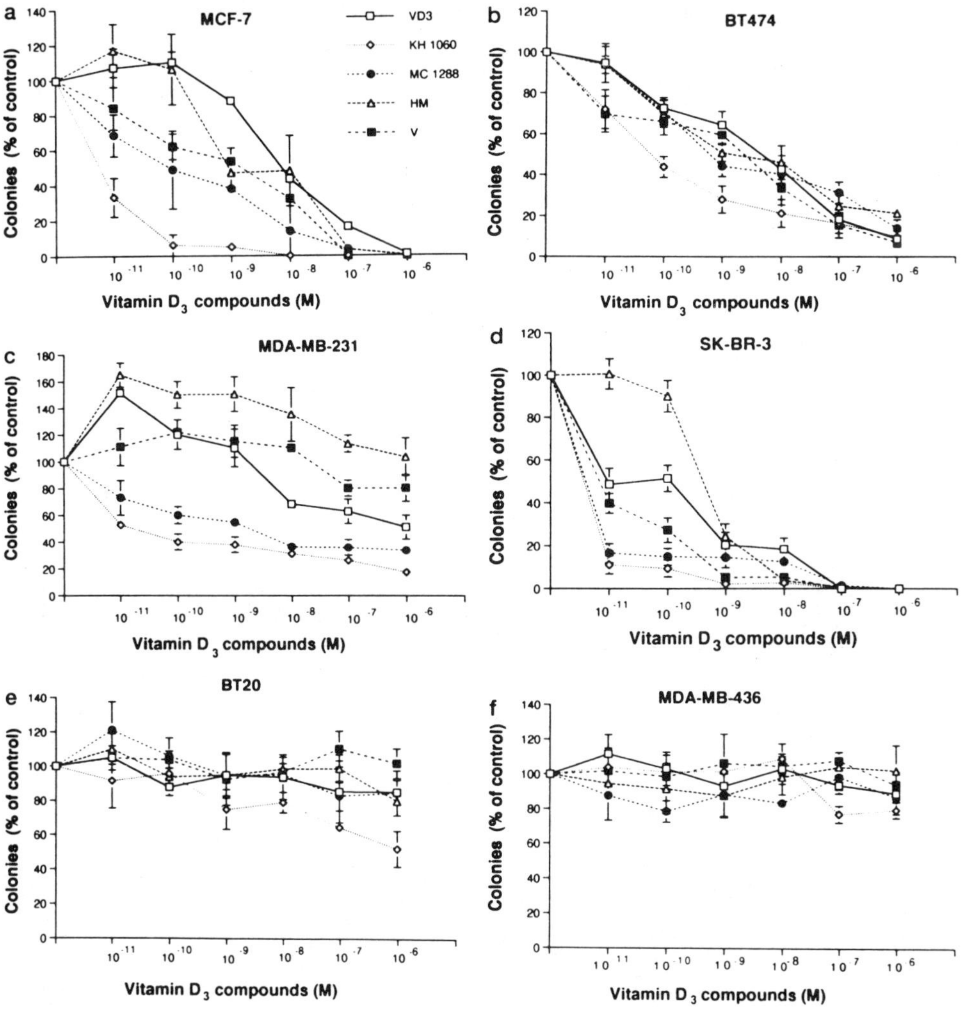

Figure 2. Dose-response effects of vitamin D_3 compounds on clonal proliferation of breast cancer cell lines (panels a - f). Results are expressed as a mean percentage ± SD of control plates containing no vitamin D_3 compounds. Each point represents a mean of at least three experiments with triplicate dishes. Control cultures ($5x10^3$/ml) contained the following colony numbers: MCF-7, 460 ± 13; BT-474, 489 ± 22; MDA- MB-231, 341 ± 17; SK-BR-3, 406 ± 19.2; BT-20, 55 ± 2; MDA-MB- 436,262±18.

(10^{-7} M) for three days, harvested with a disposable cell scraper (Costar, Cambridge, MA), pelleted in saline into cell buttons and frozen in OCT compound (Miles Diagnostics, Elkhart K. N.). Blocks were serially cryostat sectioned. These were air dried and fixed in acetone prior to immunostaining. Intracellular bcl-2 protein was detected with a murine monoclonal anti-bcl-2 antibody (DAKO, Carpinteria, CA). Cryostat sections were stained using modification of methods previously described[99,100].

The p53 was detected with monoclonal antibodies pAb 1801 (Oncogene Science Inc., Manhassett, NY) and Bp-53-12-1 (Biogenex, San Ramon, CA) using methods described previously[101]. These antibodies react with wild-type and most mutant forms of the p53 protein[102,103]. Incubation with the primary antibody (30 min.) was followed by biotinylated universal anti-immunoglobin antibodies (Shandon–Lipshaw, Pittsburg, PA) and peroxidase-conjugated streptavidin (Shandon–Lipshaw). Antibody localization for bcl-2 and p53 was performed with 3,3'-diaminobenzidene hydrochloride (Sigma, St. Louis, 5 mg/10 ml) to which 0.03% hydrogen peroxide was added just before use. Slides were counterstained with methyl green and mounted with permount. Positive controls were included with each analysis and consisted of sections of a known p53 overexpressing breast carcinoma, while negative controls consisted of substitution of the p53 specific antibody with isotype specific monoclonal antibody. Additional negative controls consisted of SKOV3 cells, a human ovarian carcinoma line that does not express p53[104]. Staining was evaluated by two investigators in a double-blinded manner.

4. RESULTS

4.1. Clonal Proliferation of Breast Cancer Cell Lines in the Presence of Vitamin D_3 Analogs

Breast cancer cells were cloned in soft agar in the presence of vitamin D_3 analogs at 10^{-11} to 10^{-6} M. Dose-response curves (Figures 2a-f) and the effective dose that inhibited 50% colony formation (ED_{50}) were determined (Table 2). The $1{,}25D_3$ analogs were effective in inhibition of clonal proliferation of 3 of the 6 breast cancer cell lines at 10^{-10}-10^{-6} M. The three sensitive lines were MCF-7, BT-474, SK-BR-3; the two resistant lines were BT-20 and MDA-MB-436. The MDA-MB-231 cell line was slightly stimulated in its clonal growth by $1{,}25D_3$ at 10^{-11}-10^{-8} M and by HM and V at 10^{-11}-10^{-9} M (Fig.2c). For the three sensitive breast cancer cell lines (Figures 2a,b,d), the KH 1060 analog [20-epi-22oxa-24a,26a,27a-tri-homo-$1{,}25(OH)_2D_3$] was the most potent compound. The second most potent $1{,}25D_3$ analog was another 20-epi-vitamin D_3 [(MC 1288) $1{,}25(OH)_2$- 20epi-D_3] which was about 100-fold more potent than $1{,}25D_3$. Compounds V and HM were effective in inhibition of clonal proliferation, but often were not much more potent than $1{,}25D_3$ (Figures 2a-f, Table 2). Those cell lines that were very resistant to $1{,}25D_3$ (BT20, MDA-MB-436) (Figures 2e, 2f) were also resistant to KH 1060 as well as to all the other vitamin D_3 analogs. The

Table 2. Inhibition of clonal proliferation of breast cancer cell lines by vitamin D_3 analogs

Breast cancer cell lines	Inhibition of clonal proliferation, ED_{50} (M)				
	VD3	KH1060	MC1288	HM	V
MCF-7	1.2×10^{-8}	8.2×10^{-12}	1.3×10^{-10}	1.0×10^{-8}	5.4×10^{-9}
BT-474	8.2×10^{-9}	8.1 x 10-11	7.4×10^{-10}	1.3×10^{-9}	4.4×10^{-9}
MDA-MB-231	1.1×10^{-6}	1.0×10^{-11}	2.2×10^{-9}	N.R.	N.R.
SK-BR-3	8.2×10^{-11}	8.0 x 10-12	1.0×10^{-11}	1.0×10^{-9}	2.4×10^{-11}
BT-20	N.R.	N.R.	N.R.	N.R.	N.R.
MDA-MB-436	N.R.	N.R.	N.R.	N.R.	N.R.

Breast cancer cells were cultured in agar for 14 days in the presence of $1{,}25D_3$ compounds (10^{-11}-10^{-6} M). Data represent three independent experiments with triplicate dishes of each concentration of analog. Cellular proliferation was determined by counting colonies. Results are expressed as a percentage of control plates containing no $1{,}25D_3$ compounds. Dose-response curves were drawn from data of Figures 2a-f and the concentration of analog achieving a 50% inhibition (ED_{50}) of clonal growth was calculated.
N.R., an ED_{50} was not reached even at 10^{-6} M of the $1{,}25D_3$ compound.

MDA-MB-231 cell line was sensitive to inhibition of proliferation only with KM1060 and MC1288 (Fig.2c).

4.2. Induction of Differentiation of Breast Cancer Cell Lines

The analysis of differentiation of breast cancer cells was determined by the expression of lipid in these cells after exposure to $1,25D_3$ and $1,25D_3$ analogs (10^{-7} M) for three days (Figures 3a-f). Approximately 10% of control cells expressed lipid. Almost all of the analogs

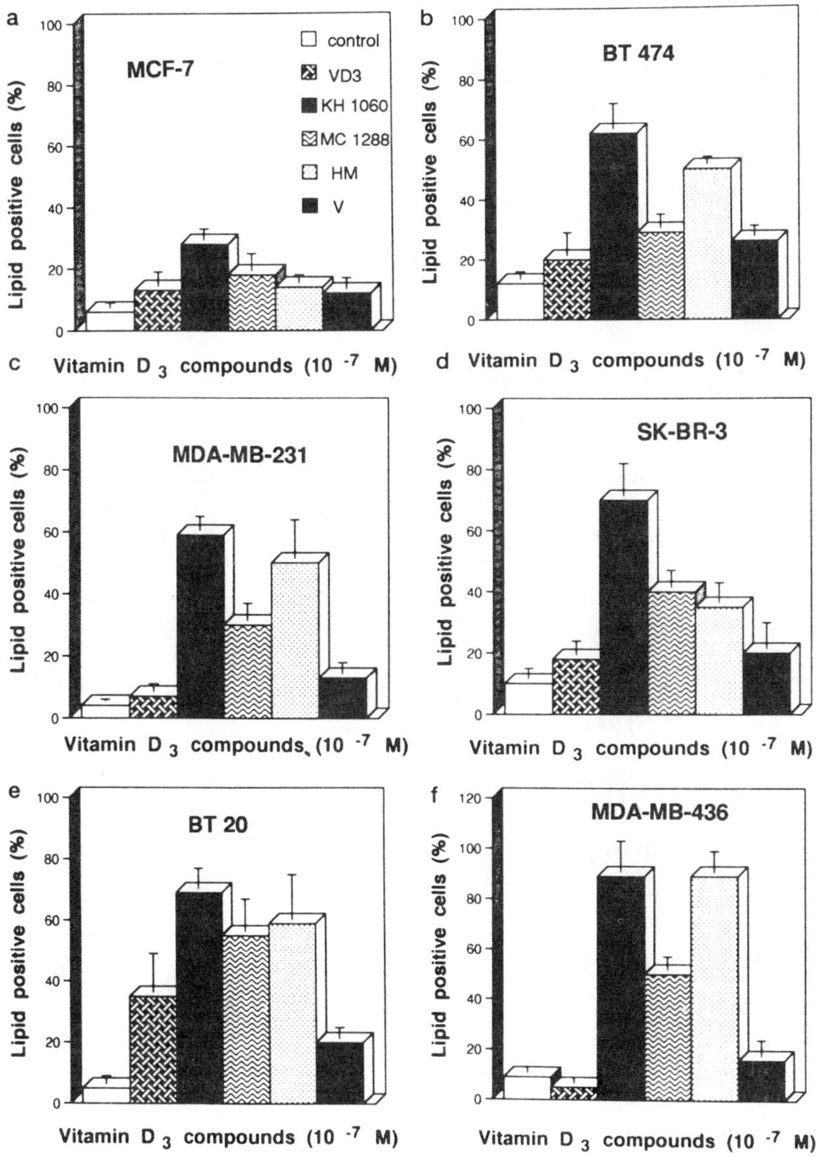

Figure 3. Induction of differentiation in breast cancer cell lines by vitamin D3 compounds. Results are expressed as a percentage of breast cancer cells that were lipid positive. Data represent means of at least three independent experiments; bars, SD.

induced lipid expression in each of the breast cancer cell lines. The most potent inducer of differentiation was KH 1060 with between 30 and 90% of the cells becoming lipid-positive in the various lines. The MC 1288 and HM were the next most potent compounds, and analog V was in general only slightly more potent than VD_3. Interestingly, the VD_3 compounds and in particular KH 1060 induced differentiation of BT20 and MDA-MB-436, which were refractory to inhibition of proliferation by VD_3 analogs. For example, 70 to 80 % of these cells became lipid positive in the presence of KH 1060 (Figures 3e-f). In contrast, these cells were resistant to the antiproliferative affects of the analogs (Figures 2e-f). This suggests a dissociation between induction of differentiation and inhibition of proliferation of these breast cancer cell lines.

4.3. Induction of Apoptosis by KH 1060 in Breast Cancer Cell Lines

The strong antiproliferative effects of KH 1060 on breast cancer cells *in vitro* may be caused by induction of apoptosis. To test this hypothesis, we examined the cell morphology and measured DNA fragmentation as markers for apoptosis. Neither morphological changes nor DNA fragmentation were detected in BT474, SK-BR-3, and MDA-MB-231 after 3, 24, 48 and 72 hours of cultivation with KH 1060 (10^{-6} M). In contrast, a mean 42±8% of MCF-7 breast cancer cells showed apoptotic morphological changes and DNA fragmentation after 48 hours of incubation with KH 1060 (10^{-6} M) (data not shown).

4.4. Decreased Levels of bcl-2 Induced by Exposure of Breast Cancer Cell Lines To KH 1060

Double-blinded analysis by immunohistochemistry showed that MCF-7, BT-474, and MDA-MB-436 cell lines had strong cytoplasmic staining (2-3+ intensity) for bcl-2; the MDA-MB-231 and BT-20 stained weakly (1+), and SK-BR-3 was negative for bcl-2 (Table 1, Figure 6). After treatment with KH-1060 (10^{-7} M for 3 days), the intensity of staining for bcl-2 decreased in MCF-7, BT474, MDA-MB-231, while little or no change in expression of bcl-2 was observed in BT-20, MDA-MB-436, and SK-BR-3 cells (Figure 6).

4.5. Nuclear Migration of p53 in MCF-7 Breast Cancer Cells After Exposure to KH 1060

The p53 protein was present predominantly in the nucleus in all the cell lines except for MCF-7 which had immunohistochemical staining for p53 mainly in the cytoplasm (Table 1). Interestingly, after exposure to KH-1060 (10^{-7} M, 3 days), the MCF-7 showed a shift in localization of p53 from the cytoplasm to the nucleus (Figure 7).

5. DISCUSSION

A focus of our investigation was the study of the biological activity and mechanism of action of novel potent vitamin D_3 analogs on breast cancer cells. These studies are important because novel non-toxic therapies are needed for the breast cancer patient who has been placed in remission but has a high likelihood of relapse and/or has a high risk of developing contralateral breast cancer.

The KH 1060 was the most active analog in the myeloid leukemia models and also in the present studies had the greatest efficacy as an inhibitor of clonal growth of the breast cancer cell lines. Remarkably, this compound was up to 4,000-fold more potent than the

Table 3. Summary of effects of KH 1060 on clonal proliferation, expression of p53 and bcl-2, induction of apoptosis, and alterations in the cell cycle in seven breast cancer cell lines

Breast cancer cell lines	Inhibition of clonal proliferation (ED_{50}, M)	Cell-cycle analysis		p53 modulation			bcl-2	Apoptosis
					Location			
		G_0/G_1	S	Intensity	n	c		
MCF-7	8.2×10^{-12}	⇑	⇓	⇒	⇑	⇓	⇓	yes
BT-474	8.1×10^{-11}	⇑	⇓	⇒	⇒	⇒	⇓	no
MDA-MB-231	1.0×10^{-11}	⇑	⇓	⇒	⇒	⇒	⇓	no
SK-BR-3	8.0×10^{-12}	⇑	⇓	⇓	⇓	⇒	-*	no
BT-20	N.R.	⇒	⇒	⇒	⇒	⇒	⇒	N.D.
MDA-MB-436	N.R.	⇒	⇒	⇒	⇒	⇒	⇒	N.D.

n-nucleus; c-cytoplasm; N.D.-none done
⇑ UP
⇓ DOWN
⇒ NO CHANGE
N.R.; ED_{50} not reached
*no bcl - 2 detactable in wild-type SK-BR-3 and levels did not change with exposure to KH 10060

physiologically active $1,25D_3$. This analog belongs to a new class of vitamin D_3 analogs which have an alteration of the stereochemistry at carbon 20 on the side chain, resulting in 20-epi-vitamin D_3 analogs. The other 20-epi-vitamin D_3 analog (MC 1288) examined in this study was also a very potent inhibitor of clonal proliferation of breast cancer cell lines (Table 2, Figures 3a-f).

The reasons for the remarkable potency of the 20-epi family of $1,25D_3$ compounds are presently unknown. The 20-epi-analogs differ markedly from $1,25D_3$ in their conformational distribution[105]. This means that the side chain of the 20-epi-vitamin D_3 analogs is directed to the left, where it is directed to the right in the "normal" isomers. This can induce significant changes in the conformation of the receptor upon ligand-binding and thereby inducing differences in the biological selectivities of various compounds.

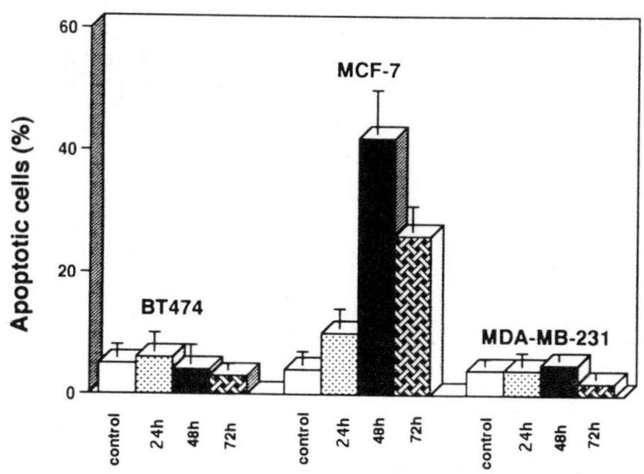

Figure 4. The ability of KH 1060 to induce apoptotic morphological changes in breast cancer cell lines. Results are expressed as a percentage of breast cancer cells that had morphological apoptotic changes as compared to control cultures that had not been exposed to VD_3 or its analogues for 24, 48, and 72h. Data represent means of at least three independent experiments; bars, SD.

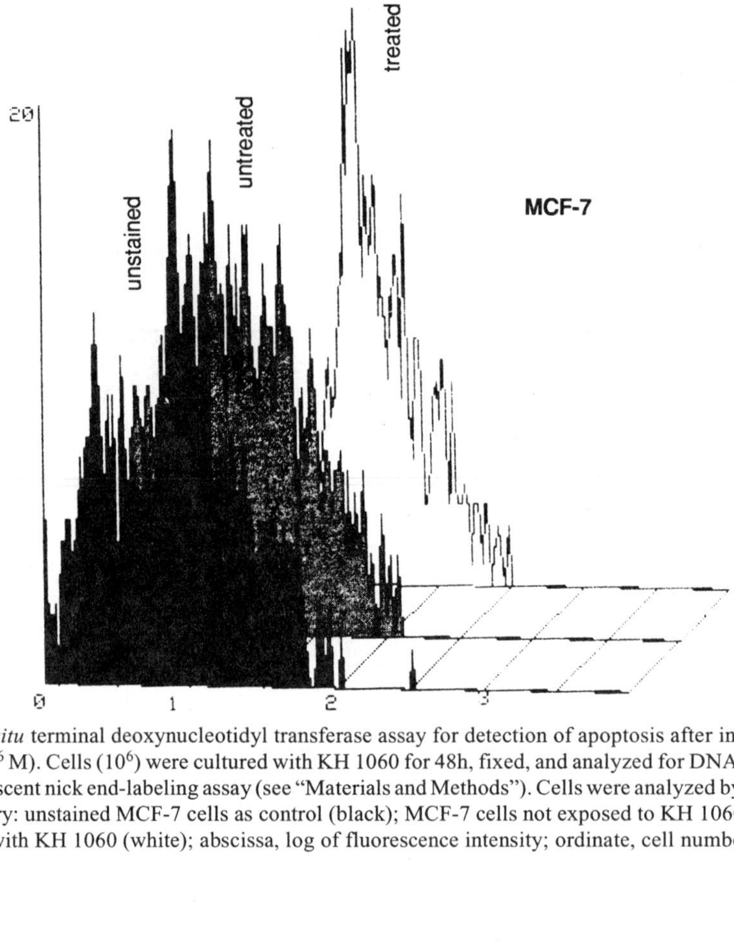

Figure 5. *In situ* terminal deoxynucleotidyl transferase assay for detection of apoptosis after incubation with KH 1060 (10^{-6} M). Cells (10^{6}) were cultured with KH 1060 for 48h, fixed, and analyzed for DNA strand breaks using a fluorescent nick end-labeling assay (see "Materials and Methods"). Cells were analyzed by fluorescence flow cytometry: unstained MCF-7 cells as control (black); MCF-7 cells not exposed to KH 1060, and MCF-7 cells treated with KH 1060 (white); abscissa, log of fluorescence intensity; ordinate, cell number.

The breast cancer cell lines in our study varied in their sensitivities to the clonal inhibitory effects of the various $1,25D_3$ compounds. The most sensitive lines, MCF-7, SK-BR-3, and BT-474 were inhibited by all the analogs. In contrast, MDA-MB-231 exhibited a slight stimulation of clonal proliferation to $1,25D_3$, HM and V over a wide range of concentrations, but showed a good response to the antiproliferative effects of the 20-epi vitamin D_3 compounds (KH 1060, MC 1288). This suggests the importance of examining these 20-epi-D_3 analogs in clinical trials and further suggests that they may have a different mode of action as compared to the other analogs. Furthermore, MDA-MB-436 and BT-20 were resistant to the antiproliferative effect of all the $1,25D_3$ compounds used in this study, suggesting that preclinical testing of the tumors *in vitro* may be helpful in selection of patients for clinical trials with these analogs.

Analysis of differentiation of breast cancer cell lines, as measured by expression of lipid markers, showed that the $1,25D_3$ compounds induced differentiation of each of the breast cancer cell lines. The most potent inducer of differentiation was KH 1060. Interestingly, KH 1060 caused the differentiation of 69 - 89% of BT20 and MDA-MB-436 cells; in contrast, these cell lines were resistant to the antiproliferative effects of this analog in clonogenic assays without any changes in their cell cycle. Also, MCF-7 cells had the weakest induction of differentiation by KH 1060 and the other $1,25D_3$ compounds; nevertheless, this cell line was very sensitive to clonal inhibition by the $1,25D_3$ and to arrest in the G_0/G_1 cell

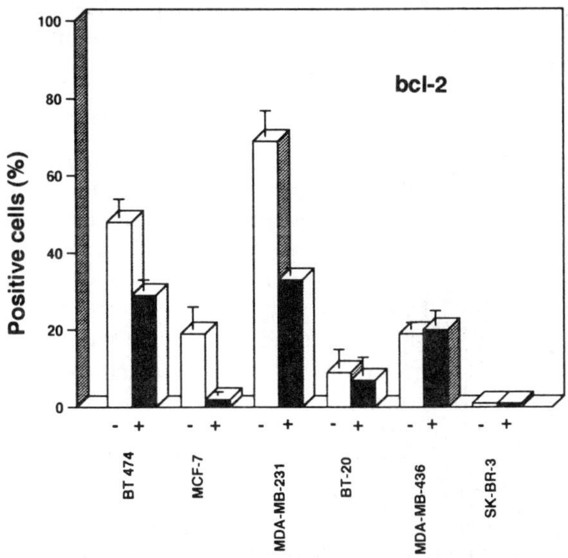

Figure 6. The effect of KH 1060 on expression of bcl-2 gene in breast cancer cell lines as measured by intensity of staining as determined by immunohistochemistry. Data represent means of at least three independent experiments, without (-) and with (+) exposure of the cells to KH 1060 (10^{-7} M for 3 days); bars, SD.

cycle after culture with KH 1060. Taken together, these results suggest that breast cancer cells can undergo at least partial phenotypic differentiation and still actively proliferate. The data suggest that regulation of proliferation and differentiation by $1,25D_3$ compounds may occur through different VDR-mediated pathways. These findings are in contrast to those of acute myeloid leukemia cells in which cessation of their proliferation and induction of their

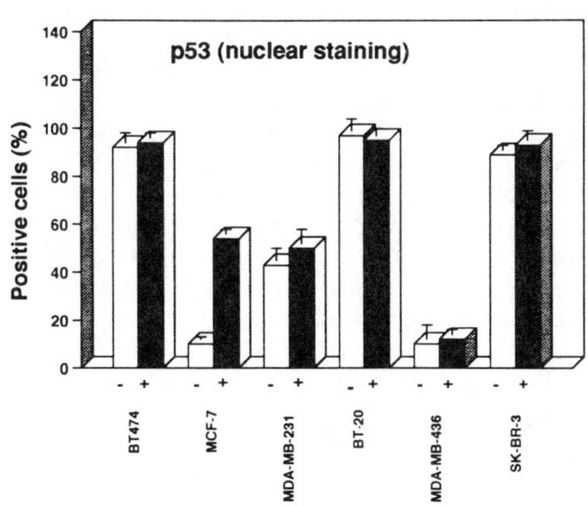

Figure 7. The effect of KH 1060 on expression of p53 in the nucleus of breast cancer cells as measured by p53 immunohistochemistry. Results are expressed as a percentage of breast cancer cells that had immunostaining for p53 in the nucleus. Data represent means of at least three independent experiments, without (-) and with (+) exposure to KH 1060 (10^{-7} M) for 3 days; bars, SD.

differentiation by the 1,25D$_3$ analogs is closely linked[106,93,107,108]. Further studies are required for identification and study of the genes that are directly modulated by these novel 1,25D$_3$ analogs and are necessary for cessation of proliferation and induction of differentiation of breast cancer cells.

Therapeutic agents can induce tumor regression through inhibition of proliferation, activation of programmed cell death (apoptosis), or both. Apoptosis is an active process and depends on expression of a specific set of genes including p53 and bcl-2[109-111]. To attempt to understand the mechanisms involved in the anti-proliferative actions of the 1,25D$_3$ analogs, the effects of KH 1060 on expression of bcl-2 and p53 and development of apoptosis were assessed in the breast cancer cell lines having different sensitivities to the antiproliferative action of this analog. The bcl-2 protein, which is over-produced in many types of human tumors, suppresses apoptosis induced by a wide-variety of stimuli, including chemotherapeutic drugs and γ-irradiation[112-115]. After incubation of the breast cancer cell lines with KH 1060, expression of the bcl-2 protein was markedly decreased in BT-474, MCF-7, and MDA-MB-231 cell lines. The same cell lines were strongly inhibited in their clonal proliferation after treatment with KH 1060. Furthermore, the BT-20 and MDA-MB-436 cells, which were refractory to the anti-proliferative effect of KH 1060, had no down-regulation of their bcl-2 after exposure to KH 1060 (Table 3), and one cell line (SK-BR-3) that expressed negligible levels of bcl-2 before and after exposure to KH 1060 was very sensitive to the growth inhibitory effects of the analog. Since all cell lines with a decreased level of expression of bcl-2 after exposure to KH 1060 or with negligible constitutive expression of bcl-2 were markedly inhibited in their clonal growth by the 1,25D$_3$ analogs, we believe that alterations of the levels of bcl-2 are directly or indirectly involved in the antiproliferative effects of the 1,25D$_3$ analogs.

Wild-type p53, a tumor suppressor protein, acts as a guardian for repair of DNA damage by preventing progression of the cell cycle from G_1 to S; this is mediated by p53 inducing the expression of p21 (WAF-1, cip-1, SDI-2) which is a cyclin-dependent kinase inhibitor[116,117]. In addition, p53 has been suggested as a key regulator of apoptosis[118-120]. The critical role played by abnormal p53 in human breast cancer is well established. About 30 to 50% of breast cancers have a mutant p53 gene and additional breast cancer samples have alteration in expression of p53[121,122]. Patients with p53 mutations in inflammatory breast carcinoma have a poor prognosis[123]; those breast cancer cell lines with mutant p53 have significantly higher levels of metastasis to the lung when transplanted subcutaneously into animals as compared to those having no p53 mutations[124]. Moll et al. have found that cells from about 30% of breast cancers have a nonfunctional wt p53 because the protein is inappropriately located in the cytoplasm of tumor cells[125]. Five of the six breast cancer cell lines utilized in our study have a mutation of the p53 gene; MCF-7 has wild-type p53. The mutated p53 protein was detected predominantly in the nucleus of these cell lines, except for MCF-7 which had p53 mainly localized in the cytoplasm. Because p53 is a nuclear transcription factor, cytoplasmic p53 probably is inactive; therefore, each of the cell lines used in this study had a non-functional p53. Furthermore, apoptosis was studied in MDA-MB-231, SK-BR-3, BT-474 and MCF-7 cells after different durations of exposure to KH 1060; no apoptosis was detected in BT-474, SK-BR-3 and MDA-MB-231. Nevertheless, KH 1060 resulted in 80-100% inhibition of their clonal proliferation and arrest in G_0/G_1, suggesting that the prominent antitumor effect of this analog in these breast cancer cell lines was independent of apoptosis. In contrast, apoptosis occurred in MCF-7 cells after 48 hours of incubation with KH 1060. Interestingly, during exposure to KH 1060, the wt p53 protein translocated from the cytoplasm to the nucleus of the MCF-7 cells, presumably resulting in a functionally active protein which could then facilitate the induction of apoptosis.

Table 3 summarizes our data concerning the effects of KH 1060 on breast cancer cell lines *in vitro*; we have identified a group of 1,25D$_3$ analogs (especially KH 1060) with potent antiproliferative and differentiative effects on breast cancer cells *in vitro*. These interesting compounds should be studied *in vivo* in appropriate breast and other cancer model systems.

REFERENCES

1. Reichel H, Koeffler HP, Norman AW: The role of the vitamin D endocrine system in health and disease. New Engl J Med 1989; 320:980-991.
2. Garland FC, Garland CF, Gorham ED, Young JF: Geographic variation in breast cancer mortality in the United States: a hypothesis involving exposure to solar radiation. Preventive Medicine 1990; 19:614-622.
3. Gorham ED, Garland FC, Garland CF: Sunlight and breast cancer incidence in the USSR. International J of Epidemiology 1990; 19:820-824.
5. Berger U, Wilson P, McClelland RA, Colston K, Haussler MR, Pike JW, Coombes RC: Immunocytochemical detection of 1,25-dihydroxyvitamin D receptors in breast cancer. Cancer Res 1987; 47:6793-6799.
6. Eisman JA, Macintyre I, Martin TJ, Frampton RJ, King RJB: Normal and malignant breast tissue is a target organ for $1,25(OH)_2$ vitamin D_3. Clin Endocrinol 1980; 13:267-272.
7. Feldman D, Chen T, Hirst M, Colston K, Karasek M, Cone C: Demonstration of 1,25-dihydroxyvitamin D_3 receptors in human skin biopsies. J Clin Endocrinol Metab 1980; 51:1463-1465.
8. Colston KW, Berger U, Coombes RC: Possible role for vitamin D in controlling breast cancer cells proliferation. Lancet 1989;28:188-191.
9. Becker AR, McDonnell DP, Hausler MR, Pike JW, Shine J, O'Malley BW: Cloning and expression of full-length cDNA encoding human vitamin D receptor. Proc Nat Acad Sci USA 1988; 85:3294-3298.
10. Minghetti PP, Norman AW: $1,25(OH)_2$-vitamin D_3 receptors: gene regulation and genetic circuitry. FASEB J 1988; 2:3043-3053.
11. Haussler MR, Mangelsdorf D, Komm BS, Terpening CM, Yamaoka K, Allegretto EA, Baker AR, Shine J, McDonnell DP, Hughes M, Weigel NL, O'Malley BW, Pike JW: Molecular biology of the vitamin D hormone. Recent Prog Horm Res 1988; 44:263-305.
12. Colston K, Colston MJ, Fieldsteel AH, Feldman D: 1,25 dihydroxyvitamin D_3 receptor in human epithelial cancer cell lines. Cancer Res 1982; 42:856-859.
13. Eisman JA, Martin TJ, MacIntyre I, Moseley JM: 1,25-dihydroxyvitamin-D_3 receptor in breast cancer cells. Lancet 1979; Dec 22/29:1335-1336.
14. Frampton RJ, Omond SA, Eisman JA: Inhibition of human cancer cell growth by 1,25-dihydroxyvitamin D_3 metabolites. Cancer Res 1983; 43:443-447.
15. Abe E, Miyaura C, Sakagami H, Takeda M, Konno K, Yamazaki T, Yoshiki S, Suda T: Differentiation of mouse myeloid leukemia cells induced by 1,25-dihydroxyvitamin D_3. Proc Natl Acad Sci USA 1981; 76:4990-4994.
16. Colston KW, Chander SK, Mackay AG, Coomes RC: Effects of synthetic vitamin D analogues on breast cancer cell proliferation *in vivo* and *in vitro*. Biochem Pharmacology 1992; 44:693-702.
17. Danielpour D, Kadomatsu K, Anzano MA, Smith JM, Sporn MB: Development and characterization of nontumorigenic and tumorigenic epithelial cell lines from rat dorsal-lateral prostate. Cancer Res 1994; 54:3413-21.
18. Bower M, Colston KW, Stein RC, Hedley A, Gazet JC, Ford HT, Coombes RC: Topical calcipotriol treatment in advanced breast cancer. Lancet 1991; 337:701-702.
19. Eisman JA, Barkla DH, Tutton PJM: Suppression of *in vivo* growth of human cancer solid tumor xenografts by $1,25(OH)_2$ vitamin D_3. Cancer Research 1987; 47:21-25.
20. Honma Y, Hozumi M, Abe E, Konno K, Fukushima M, Hata S, Nishil Y, DeLuca HF, Suda T: 1-α-hydroxyvitamin D_3 prolong survival time of mice inoculated with myeloid leukemia cells. Proc Acad Sci USA 1983; 80:201-204.
21. Sato T, Takusagawa K, Asoo N, Konno K: Effects of 1-αhydroxyvitamin D_3 on metastasis of rat ascites hepatoma K-231. Br J Cancer 1984; 50:123-125.
22. Zhou JY, Norman AW, Chen DL, Sun GW, Uskokovic M, Koeffler HP: 1,25-Dihydroxy-16-ene-23-yne-vitamin D_3 prolong survival time of leukemic mice. Proc Natl Sci USA 1990; 87:3929-3932.
23. Frappart L, Falette N, Lefebvre M, Bremond A, Vauzelle JL, Saez S: *in vitro* study of effects of 1,25 dihydroxyvitamin D_3 on the morphology of human breast cancer cell line BT 20. Differentiation 1989; 40:63-69.

24. Binderup L, Latini S, Binderup E, Bretting C, Calverley M, Hansen K: 20-epi-vitamin D_3 analogues: A novel class of potent regulators of cell growth and immune responses. Biochem Pharmac 1991; 42:1569-1575.
25. Mathiasen IS, Colston KW, Binderup L: EB 1089, a novel vitamin D analogue, has strong antiproliferative and differentiation inducing effects on cancer cells. J Steroid Biochem Molec Biol 1993; 46:365-371.
26. Dokoh S, Donaldson CA, Haussler MR: Influence of 1,25 dihydroxyvitamin D_3 on cultured osteogenic sarcoma cells; connection with 1,25 dihydroxyvitamin D_3 receptor. Cancer Res 1984; 44:2103-2109.
27. Miyaura C, Abe E, Kuribayashi T, Tauka H, Konno K, Nishil Y, Suda T: 1,25-dihydroxyvitamin D_3 induces differentiation of human myeloid leukemia cells. Biochem Biophys Res Common 1981; 102:937-943.
28. Brehier A, Thomasset M: Human colon cell line HT-29. Characterization of 1,25-dihydroxyvitamin D_3 receptor and induction of differentiation by the hormone. J Steroid Biochem 1988; 29:265-270.
29. Norman AW, Zhou JH, Henry HL, Uskokovic MR, Koeffler HP: Structure-function studies on analogues of 1-α,25-dihydroxyvitamin D_3: differential effects on leukemic cell growth, differentiation, and intestinal absorption. Cancer Res 1990; 50:6857-6864.
30. Kragballe K, Wildfang I: Calcipotriol (MC903), a novel vitamin D_3 analogue, stimulates terminal differentiation and inhibits proliferation of cultured human keratinocytes. Dermatol Res 1990; 282:164-167.
31. Mangelsdorf DJ, Koeffler HP, Donaldson CA, Pike JW, Haussler MR: Induced differentiation in a human promyelocytilc leukemia cell line (HL-60): receptor-mediated maturation to macrophage-like cells. J Cell Biol 1984; 98:391-398.
32. Rigby WFC, Shen L, Ball ED, Juyre PM, Fanger FW: Differentiation of a human monocytic cell line by 1,25dihydroxyvitamin D_3 (calcitrol): A morphologic, phenotypic, and functional analysis. Blood 1984; 64:1110-1115.
33. McCarthy DJ, Hibbin JF, San Miguel JF, Freake HC, Rodriguez B, Andrews C, Pining A, Catovsky D, Goldman JM: The effect of vitamin D_3 metabolites on normal and leukemic bone marrow cells *in vitro*. Int J Cell Cloning 1984; 2:227-242.
34. Koeffler HP, Armatruda T, Ikekawa N, Kobayashi Y, DeLuca HF: Induction of macrophage differentiation of human normal and leukemic myeloid stem cells by 1,25-dihydroxyvitamin D_3 and its fluorinated analogues. Cancer Res 1984; 44:5624-5628.
35. Koeffler HP: Study of normal and abnormal hematopoiesis using human leukemic cell lines. UCLA Symposia on Molecular and Cellular Biology. New Series, Vol.9. New York, New York: Alan R Liss, Inc 1983; 179-195.
36. Reichel H, Koeffler HP, Barbers R, Munker R, Norman A: 1,25 dihydroxyvitamin D_3 and the hematopoietic system. In: Norman AW, Schaefer K, Grigoleit HG, Herrath DV eds. Vitamin D: Chemical, biochemical and Clinical Update. Berlin, Germany: Walter de Gruyter 1985; 167-176.
37. Reichel H, Koeffler HP, Bishop JE, Norman AW: Effects of lipopolysaccharides on conversion of 25-hydroxyvitamin D_3 by normal human macrophages. In:Norman AW, Schaefer K, Grigoleit HG, Herrath DV eds. Vitamin D:Chemical, Biochemical and Clinical Update. Berlin, Germany: Walter de Gruyter 1985; 61-62.
38. Bishop JE, Reichel H, Koeffler HP, Norman AW: $1,25(OH)_2D_3$ receptor in pulmonary alveolar macrophages. In: Norman AW, Schaefer K, Grigoleit HG, Herrath DV eds. Vitamin D: Chemical, Biochemical and Clinical Update. Berlin, Germany:Walter de Gruyter 1985; 127-128.
39. Koeffler HP, Reichel H, Munker R, Barbers R, Norman A: Interaction of 1,25-Dihydroxyvitamin D_3 and the hematopoietic systems. In: Gale RP and Golde DW eds. Leukemia: Recent Advances in Biology and Treatment. Vol.28. New York, New York: Alan R. Liss, Inc. 1985; 399-413.
40. Koeffler HP: Study of differentiation and proliferation of leukemic cells using myeloid leukemia lines. In: Bloomfield CD ed. Adult Leukemias 2. The Hague, Netherlands: Martinus Nijhoff Publishers 1985; 27-68.
41. Koeffler HP: Vitamin D: Myeloid differentiation and proliferation. Current Trends in Human Leukemia VI. In: Gallo R, Neth H, Moore M. eds. Hematology and Blood Transfusion Vol.29. Modern Trends in Human Leukemia. Berlin, Germany: Springer-Verlag 1985; 409-417.
42. Reichel H, Koeffler HP, Bishop JE, Norman AW: Interactions between lymphocytes and macrophages mediated by 1,25-dihydroxyvitamin D_3 and τ-interferon. In: Cohen DV, Martin TV, Meunier PJ, eds. Calcium Regulation and Bone Metabolism: Basic and clinical aspects. Vol 9. amsterdam, Holland: Excerpta Medica 1987; 371-376.
43. Koeffler HP, Tobler A, Reichel H, Norman AW: Interaction of 1,25-dihydroxyvitamin D_3 with normal and abnormal hematopoiesis. In: Aarbakke J, Chiang PK, Koeffler HP eds. Tumor Cell Differentiation, Biology and Pharmacology. Clifton, New Jersey: The Humana Press 1987; 137-157.

44. Lubbert M, Koeffler HP: Leukemic cell lines as a paradigm for myeloid differentiation. In: Olsson, L. ed. in chief. Cancer Reviews, Vol 10. Copenhagen, Denmark: Munksgaard International Publishers, Ltd 1988; 33-62.
45. Collins ED, Jones GI, Uskokovic M, Koeffler HP, Norman AW: 1,25$(OH)_2$ vitamin D_3 analogs: Comparison of 1,25$(OH)_2D_3$ receptor ligand specificity and biological activity in the chick and human promyelocytic (HL-60) cells. In: Norman AW, Schaefer K, Grigoleit HG, Herrath DV eds. Vitamin D:Molecular, Cellular and clinical Endocrinology. Berlin, West Germany: Walter de Gruyter 1988; 433-434.
46. Paquette R, Koeffler HP: Retinoids and 1,25 dihydroxyvitamin D_3: Role in myelodysplastic syndromes and leukemia. In: Proceedings of International Symposium on Myelodysplastic Syndromes: Pathophysiology and Treatment Kyoto, Japan. The Netherlands: Elsevier Science Publishers 1988; 173-186.
47. Paquette RL, Koeffler HP: Myelodysplastic Syndrome: Culture Studies and Therapy. In: Mufti GJ and Glaton DAG eds. Churchill Livingstone Monograph of Myelodysplastic Syndromes. Amsterdam, The Netherlands: Elsevier Science Publishers 1992.
48. Koeffler HP, Norman AW: Interaction between the hematopoietic system and the vitamin D endocrine system. In: Norman AW, Schaefer K, Grigoleit HG, Herrath DV eds. Vitamin D: Molecular, cellular and clinical endocrinology. Berlin: Walter De Gruyther 1988; 263-266.
49. Koeffler HP, Reichel H, Tobler A, Norman AW: Macrophages and vitamin D_3. In: Ahserson GL, Zembala M. eds. The Human Monocyte. London, England: Academic Press Limited 1989; 345-351.
50. Miller CW, Koeffler HP: *In vivo* and *in vitro* differentiation of myeloid leukemic cells. In: Mauer AM, ed. The biology of Human Leukemia. Baltimore, Maryland: Johns Hopkins Press 1990; 152-176.
51. Reichel H, Koeffler HP, Norman AW: Production of 1α, 25-dihydroxyvitamin D_3 in hematopoietic cells. In: Peterlik M, Bronner F. eds. Molecular and cellular regulation of calcium and phosphate metabolism. New York, New York: Alan R. Liss, Inc. 1990; 81-97.
52. Kizaki M, Koeffler HP: 1,25-dihydroxyvitamin D_3 and hematopoietic cells. In: Vitamins and Cancer Prevention. Laidlaw SA, Swendseid ME eds. New York, New York: Wiley and Liss. Inc. 1991; 91-102.
53. Jordan SC, Sakai R, Koeffler HP, Lemire J, Adams JS: Immunoregulatory and Prodifferentiating effects of 1,25-dihydroxyvitamin D_3 in human mononuclear cells. Nutrient Modulation of the Immune response. Marcel Dekker, Inc., 1992; 3-29.
54. Kizaki M, Koeffler HP: Differentiation inducing agents in the treatment of MDS. Seminars in Oncology, Myelodysplastic Syndromes 1992; 95-105.
55. Tobler A, Koeffler HP: Differentiation inducers and their potential use in the treatment of acute myelogenous leukemia (MJ Waring and BAJ Ponder, ed.) Kluwer Academic Publishers 1991; 163-183.
56. Paquette R, Koeffler HP: Differentiation therapy for myelodysplastic syndromes. North American Clinics of Hematology (HP Koeffler, ed.), W.B. Sunders Company 1992; 687-706.
57. Akashi M, Koeffler HP: Differentiation therapy in leukemia. Focus on CML/MM. KK Standard McIntyre Co. Tokyo, Japan. 1994; 5-7.
58. Elstner E, Dawson MI, de Vos S, Pakkala S, Binderup L, Okamura W, Uskokovic M, Koeffler HP: Myeloid differentiation mediated through new potent retinoids and vitamin D_3 analogs. In: Hematology and blood Transfusion, Buechaer, Hiddenmann (eds), Springer-Verlag, accepted, 1994.
59. Koeffler HP: Induction of differentiation of human acute myelogenous leukemia cells: Therapeutic implications. Blood 1983;62:709-721.
60. Koeffler HP, Hiriji K, Itri L: 1,25-dihydroxyvitamin D_3: *In vivo* and *in vitro* effects on human preleukemic and leukemic cells. Cancer Treatment Rep 1985; 69(12):1399-1407.
61. Koeffler HP, Reichel H, Bishop, Norman AW: γ-interferon stimulates production of 1,25-dihydroxyvitamin D_3 by human pulmonary macrophages. Biochem Biophys Res Commun 1985; 127:596-603.
62. Koeffler HP, Yelton L, Prokocimer M, Hirji K: Study of differentiation of fresh myelogenous leukemic cells by compounds that induce a human promyelocytic leukemic line (HL-60) to differentiate. Leukemia Res 1985; 9:73-81.
63. Munker R, Norman A, Koeffler HP: Vitamin D compounds: Effect on clonal proliferation and differentiation of human myeloid cells. J Clin Invest 1986; 78:474-480.
64. Reichel H, Koeffler HP, Norman A: Regulation of 25-hydroxyvitamin D_3 metabolism in a human promyelocytic leukemia cell line (HL-60): 1,25 dihydroxyvitamin D_3 stimulates the synthesis of 24,25-dihydroxyvitamin D_3. Arch Biochem Biophys 1986; 251:222-231.
65. Munker M, Munker R, Saxton RE, Koeffler HP: Effect of recombinant monokines, lymphokines and other agents on clonal proliferation of cells from human lung cancer lines. Cancer Res 1987; 47:4081-4085.
66. Reichel H, Koeffler HP, Tobler A, Norman AW: 1,25-dihydroxyvitamin D_3 inhibits γ- interferon synthesis by normal human peripheral blood lymphocytes. Proc Natl Acad Sci USA 1987; 84:3385-3389.

67. Reichel H, Koeffler HP, Norman AW: Synthesis *in vitro* of 1,25- dihydroxyvitamin D_3 and 24,25 dihydroxyvitamin D_3 by interferon gamma stimulated normal human and alveolar bone marrow macrophages. J Bio Chem 1987; 262(23):10931-10937.
68. Reichel H, Koeffler HP, Norman AW: 25-hydroxyvitamin D_3 metabolism by human T- lymphotropic virus transformed lymphocytes. J Clin Endocr Metab 1987; 5:519-526.
69. Reichel H, Koeffler HP, Bishop JE, Norman AW: 25-hydroxyvitamin D_3: metabolism by lipopolysaccharide-stimulated normal human macrophages. J Clin Endoc and Metab 1987; 64:1-9.
70. Reichel H, Koeffler HP, Barbers R, Norman AW: Regulation of 1,25-dihydroxyvitamin D_3 production by cultured alveolar macrophages from normal human donors and from patients with pulmonary sarcoidosis. J Clin Endocr Metab 1987; 65:1201-1209.
71. Tobler A, Koeffler HP: Differentiation of human acute myelogenous leukemia cells: Therapeutic possibilities. Acta Haematologica 1987; (78)1:127-135.
72. Tobler A, Gasson J, Reichel H, Norman AW, Koeffler HP: Granulocyte-macrophage colony stimulating factor: sensitive and receptor-mediated regulation by 1,25-dihydroxyvitamin D_3 in normal human peripheral blood lymphocytes. J Clin Invest 1987; 79:1700-1705.
73. Lubbert M, Koeffler HP: Myeloid cell lines: Tools for studying differentiation of normal and abnormal hematopoietic cells. Blood Reviews 1988; 2:121-133.
74. Tobler A, Miller C, Norman AW, Koeffler HP: 1,25-dihydroxyvitamin D_3 modulates the expression of a lymphokine (granulocyte-monocyte colony stimulating factor) post-transcriptionally. J Clin Invest 1988; 81:1819-1823.
75. Zhou JY, Norman AW, Lubbert M, Collins ED, Uskokovic MR, Koeffler HP: Novel vitamin D analogs that modulate leukemic cell growth and differentiation with little effect on either intestinal calcium absorption or bone calcium mobilization. Blood 1989; 74:82-93.
76. Koeffler HP, Bishop JE, Reichel H, Singer F, Nagler A, Tobler A, Walka W, Norman AW: Lymphocyte cell lines from vitamin D-dependent rickets type II show functional defects in the 1,25 dihydroxyvitamin D_3 receptor. Mol and Cellul Endo 1990; 70:1-11.
77. Reichel H, Koeffler HP, Norman AW: Production of 1 alpha, 25 dihydroxyvitamin D_3 by hematopoietic cells. Progress in Clinical and Biological Research 1990; 332:81-97.
78. Norman AW, Zhou JY, Henry HL, Uskokovic MR, Koeffler HP: Structure-function studies on analogs of 1α,25-dihydroxyvitamin D_3: Differential effects on leukemic cell growth, differentiation, and intestinal calcium absorption. Cancer Research 1990; 50:6857-6864.
79. Kizaki M, Norman AW, Bishop JE, Lin CW, Karmakar A, Koeffler HP: 1,25-dihydroxyvitamin D_3 receptor RNA: expression in hematopoietic cells. Blood 1991;77:1238-1247.
80. Zhou JY, Norman AW, Akashi M, Chen DL, Uskokovic M, Aurrecocchea JM, Dauben WG, Okamura W, Koeffler HP: Development of a novel $1,25(OH)_2$-Vitamin D_3 analog with potent ability to induce HL-60 cell differentiation without modulating calcium metabolism. Blood 1991; 78:75-82.
81. Figadere B, Okamura WH, Norman AW, Henry HL, Koeffler HP, Zhou JY: Arocalciferols: Synthesis and biological evaluation of aromatic side chain analogues of 1α,25-dihydroxyvitamin D_3. J Med Chem 1991; 34:2452-2463.
82. Kizaki M, Koeffler HP: differentiation-inducing agents in the treatment of myelodysplastic syndromes. Seminars in Oncology 1992; 19:95-105.
83. Reichel H, Bishop JE, Koeffler HP, Norman AW: Evidence for 1,25-dihydroxyvitamin D_3 production by cultured porcine alveolar macrophages. Mol Cell Endo 1991; 75:163-167.
84. Norman AW, Koeffler HP, Bishop JE, Collins ED, Sergeev I, Zhou LX, Nemere I, Zhou J, Henry HL, Okamura: Structure-Function relationships in the vitamin D endocrine system for $1,25(OH)_2D_3$ analogs. Proceedings of the Eighth Workshop on Vitamin D 1991; 146-154.
85. Bishop JE, Henry HL, Okamura WH, Koeffler HP, Norman A: Acrocalciferols: New class of side-chain analogs of $1,25(OH)_2D_3$ vitamin gene. Proceedings of the Eighth Workshop on Vitamin D 1991; 165-166.
86. Kizaki M, Simon KJ, Ikeda Y, Nanjo M, Koeffler HP: Effect of 1,25-dihydroxyvitamin D_3 and its analogs on human immunodeficiency virus infection in monocytes/macrophages. Leukemia 1993; 7(10):1525-1530.
87. Jung SJ, Lee YY, Pakkala S, de Vos S, Elstner E, Norman AW, Green J, Uskokovic M, Koeffler HP: $1,25(OH)_2$-16ene-vitamin D_3 is a potent antileukemic agent with low potential to cause hypercalcemia. Leukemia Research 1994; 18(6):453-463.
88. Elstner E, Lee YY, Hashiya M, Pakkala S, Binderup L, Norman AW, Okamura WH, Koeffler HP: 1α,25-dihydroxy-20-epi-vitamin D_3: an extraordinarily potent inhibitor of leukemic cell growth *in vitro*. Blood 1994; 84(6):1960-1967.

89. Pakkala S, de Vos S, Elstner E, Rude RK, Uskokovic M, Binderup L, Koeffler HP: Vitamin D_3 analogs: effect on leukemic clonal growth and differentiation, and on serum calcium levels. Leukemia Research 1995; 19:65-72.
90. Imai Y, Pike JW, Koeffler HP: Potent vitamin D_3 analogs: their abilities to enhance transactivation and to bind to the vitamin D_3 response element. Leukemia Research, in press, 1994.
91. Elstner E, Said J, Umiel T, de Vos S, Israeli ML, Heber D, Binderup L, Uskokovic M, Koeffler HP: 20-epi-vitamin D_3 analogs: a novel class of potent inhibitors of proliferation and inducers of differentiation in human estrogen receptor (ER)-positive and ER-negative breast cancer cell lines *in vitro*. Cancer Research, in press, 1995.
92. Ostrem VI, Tanaka Y, Prahl J, DeLuca HF, Ikekawa N: 24- and 26-homo 1,25-dihydroxyvitamin D_3. Preferential activity in inducing differential of human leukemia cell HL-60 *in vitro*. Proc Natl Acad Sci USA 1987; 84:2610.
93. Elstner, E., Lee, Y. Y., Hashiya, M., Pakkala, S., Binderup, L., Norman, A. W., Okamura, W. H., and Koeffler H. P.: 1α,25-dihydroxy-20-epi-vitamin D_3: an extraordinarily potent inhibitor of leukemic cell growth *in vitro*. Blood, 84:1968-1968, 1994.
94. Kiguchi, K., Giometti, C., Chubb, C. H., Fujiki, H., and Huberman, E.: Differentiation induction in human breast tumor cells by okadaic acid and related inhibitors of protein phosphatases 1 and 2A. Biochem and Biophysical Research Communications, 189:1261-1267, 1992.
95. Sheehan, D. C.: Theory and practice of histotechnology, 2nd ed C. V. Mosby Company. St. Lous, p.209, 1980.
96. Bacus, S. S., Kiguchi, K., Chin, D., King, C. R., and Huberman, E. Differentiation of cultured human breast cancer cells (AU-565 and MCF-7) associated with loss of cell surface HER-2/neu antigen. Molecular Carcinogenesis, 3: 350-362, 1990.
97. Martin, S. J. and Cotter T. G. Apoptosis of human leukemia: induction, morphology, and molecular mechanisms. In: Apoptosis II: Molecular basis of apoptosis in disease. Cold Spring Harbor Laboratory Press, p.1, 1994.
98. Sellins, K. S., and Cohen, J. J.: Gene induction by Gamma-irradiation leads to DNA fragmentation in lymphocytes. J. Immunology, 139: 3199-3206, 1987.
99. Said, J. W., Pinkus, G. S., Lones, M. A., Preston, M., and Shintaku, I. P. Expression of bcl-2 oncogene protein in malignant lymphomas, including Hodgkin's disease and non-Hodgkin's lymphomas of T-cell phenotype. Applied Immunohistochemistry, 1: 108-114, 1993.
100. Lones, M. A., Said, J. W., Pinkus, G. S. Bcl-2 oncogene protein in preferentially expressed in Reed Sternberg cells in Hodgkin's disease of modular sclerosis subtype. Am J. Clin Pathol. (in press), 1994.
101. Walts, A. E., Said, J. W., and Koeffler, H. P. Is immunoreactivity for p53 useful in distinguishing benign from malignant effusions? Localization of p53 gene product in benign mesothelial and adenocarcinoma cells. Modern Pathol. (in press), 1994.
102. Banks, L., Matlashewski, G., and Crawford, L. Isolation of human p53 specific monoclonal antibodies and their use in the studies of human p53 expression. Eur. J. Biochem., 159: 529-534, 1986.
103. Bartck,J. Iggo, R., Gannon, J., and Lane, D. P. Genetic and immunochemical analysis of mutant p53 in human breast cancer cell lines. Oncogene, 5: 893-9, 1990.
104. Miller, C. W., Chumakov, A., Said, J., Chen, D. L., Aslo, A., and Koeffler, H. P. Mutant p53 proteins have diverse intracellular abilities to oligomerize and activate transcription. Oncogene 8: 1815-1824, 1993.
105. Midland, M.M., Plumet, Y., and Okamura, W.H. Effect of C20 stereochemistry on the conformational profile of the side chains of vitamin D analogs. Bioorganic and Medicinal Chemistry Letters, 3: 1799-1804, 1993.
106. Mangelsdorf, DJ, Koeffler HP, Donaldson CA, Pike JW, Haussler MR: Induced differentiation in a human promyelocytylic leukemia cell line (HL 60): receptor-mediated maturation to macrophage-like cells. J Cell Biol 1984; 98:391-398.
107. Collins, S. J., Ruscetti, F. W., Gallagher, R. E., and Gallo, R. C. Normal functional characteristics of cultured human promyelocytic leukemia cells (HL-60) after induction of differentiation by dimethylsulfoxide. J. Exp. Med., 149: 969-977, 1979.
108. Douer, D., and Koeffler, H. P. Inhibition of the clonal growth of human myeloid leukemia cells. J. Clin. Invest. , 69: 277-283.
109. Fisher, D. E. Apoptosis in cancer therapy: crossing the threshold. Cell, 78: 539-542.
110. Raff, M. C. Social control on cell survival and cell death. Nature (Lond.), 356: 397 400, 1992.
111. Oltval, Z. N., and Korsmeyer, S. J. Checkpoints of dueling dimers foil death wishes. Cell, 79: 189-192, 1994.
112. Vaux,, D. L., Cory, S., and Adams, J. M. Bcl-2 gene promotes haemopoietic cell survival and cooperates with c-myc to immortalize pre-B cells. Nature,335: 440-442, 1988.

113. Hockenbery, D., Nu–ez, G., Milliman, C., Schreiber, R. D., and Korsmeyer, S. J. Bcl- 2 is an inner mitochondrial membrane protein that blocks programmed cell death. Nature, 348: 334-339, 1990.
114. Hockenbery, D. M. The bcl-2 oncogene and apoptosis. Seminars in Immunology, 4: 413-420, 1992.
115. McDonnell, T. J., Deane, N., Platt, F.M., Nunez, G., Jaeger, U., McKearn, J. P., and Korsmeyer, S. J. Bcl-2 immunoglobin transgenic mice demonstrate extended B cell survival and follicular lymphoproliferation. Cell, 57: 79-88, 1989.
116. El—Deiry, W. S., Tokino, T, Velculescu, V. E., Levy, D. B., Parsons, R., Trent, J. M., Lin, D., Mercer, W. E., Kinzler, W. E., and Vogelstein, B. WAF1, a potential mediator of p53 tumor suppression. Cell, 75: 817-825, 1993.
117. Herper, J. W., Adam G. R., Wei, N, Keyomarsi, K., and Elledge, S. J. The p21 cdk- interacting protein Cip 1 is a potent inhibitor of G1 cyclin-dependent kinases. Cell 75: 805-816, 1993.
118. Symonds, H., Krall, L., Remington, L., Saenz-Robles, L., Lowe, S., Jacks, T., and Van Dyke, T. p53-dependent apoptosis suppresse tumor growth and progression *in vivo*. Cell, 78: 703-711, 1994.
119. Yonish-Rouach, E., Resnitzky, D., Lotem, J., Sachs, L., Kimchi, A., and Oren, M. Wild-type p53 induces apoptosis of myeloid leukemic cells that is inhibited by interleukin- 6. Nature (Lond.), 352: 345-347, 1991.
120. Show, P., Bovey, R., Tardy, S., Sahli, R., Sordat, B., and Costa, J. Induction of apoptosis by wild-type p53 in a human colon tumor-derived cell line. Proc. Natl. Acad. Sci. USA, 89: 4495-4499, 1992.
121. Crawford,L. V., Pim, D. C., And Lamb, P. The cellular protein p53 in human tumours. Mol. Biol. Med., 2: 261-272, 1984.
122. Cattoretti, G., Rilke, F., Andreola, S., D'Amato, L., and Domenico, D. p53 in breast cancer. Int. J. Cancer, 41: 178-183, 1988.
123. Riou, G., L, M. G., Travagli, J. P., Moll, U. M., and Levine, A. J. Poor prognosis of p53 nuclear overexpression and mutation in inflammatory breast carcinoma. Proc. Amer. Assoc. Canc. Res., 35: Abstract 1314, 1994.
124. Mehta, R. R., Graves, J. M., and Das Gupta, T. K. p53 levels and breast tumor metastasis in experimental animals. Proc. Amer. Assoc. Canc. Res., 35: Abstract 286, 1994.
125. Moll,U. M., Riou, G., and Levine, J. Two distinct mechanisms alter p53 in breast cancer: Mutation and nuclear exclusion. Proc. Natl. acad. Sci. USA, 89: 7262-7266, 1992.

6

NOVEL LIPIDS AND CANCER

Isoprenoids and Other Phytochemicals

Charles E. Elson

Department of Nutritional Sciences
University of Wisconsin-Madison
1415 Linden Drive
Madison, Wisconsin 53706

1. INTRODUCTION

The focus of studies of diet/cancer interrelationships is shifting from the roles nutrients play to the chemopreventive actions of plant-derived, non-nutrient dietary constituents (indoles, coumarins, phthalides, organosulfur compounds, phytoestrogens, isothiocyanates, bioflavonoids and protease inhibitors). Our interest in this aspect of diet/health relationships has roots in findings that products of plant mevalonate pathways both suppress hepatic mevalonate synthesis with a concomitant lowering of serum LDL cholesterol level and suppress the growth of chemically-initiated and transplanted tumors.[1-4] Some of the more effective end products derived from the sequential mevalonate pathway intermediates, geranyl pyrophosphate, farnesyl pyrophosphate and geranylgeranyl pyrophosphate, are listed on Figure 1.[1, 2] The pathway provides essential dietary components, specifically phylloquinone, β-carotene and α-tocopherol. The monoterpenes and sesquiterpenes have no nutritional impact. Never the less each of the products listed on Figure 1 has been shown to have an impact on HMG CoA reductase activity, blood cholesterol level and tumor cell proliferation.[1-4]

The isoprenoids emphasized in this presentation, β-ionone and γ-tocotrienol, are related to β-carotene and α-tocopherol, the "antioxidant vitamins" linked by epidemiological data[5], but not by randomized experimental trials[6], with a reduced incidence of cancer. β-Ionone has no antioxidant activity and that of γ-tocotrienol is marginal; both are potent suppressors of hepatic mevalonate synthesis. We postulate that the isoprenoid-mediated suppression of mevalonate synthesis diminishes the pool of mevalonate pathway intermediates which are required cholesterol synthesis and for the post translational modification of proteins essential for cell survival (Figure 2). We further postulate that the isoprenoid-mediated suppression of tumor growth traces to the restrictions this action imposes on the post translational modification of nuclear lamins and ras oncoproteins (Figure 2). As shown on Figure 2 the mevalonate pathway offers targets for controlling cholesterol synthesis and tumor growth with lovastatin and mevinolin[7], sodium phenylacetate[8] and fluoromevalonate.[9]

Dietary Fats, Lipids, Hormones, and Tumorigenesis, edited by Heber and Kritchevsky
Plenum Press, New York, 1996

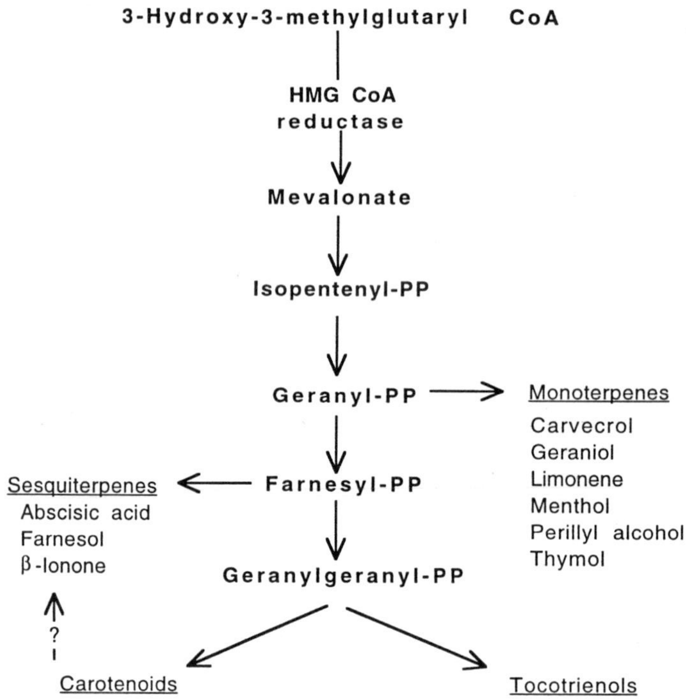

Figure 1. Intermediate and end products of mevalonate metabolism in plant tissues. HMG CoA reductase is the rate-limiting activity for the synthesis of mevalonic acid[14] the rate-limiting substrate for the synthesis of the isoprenoid products shown to have tumor-suppressive actions.[2]

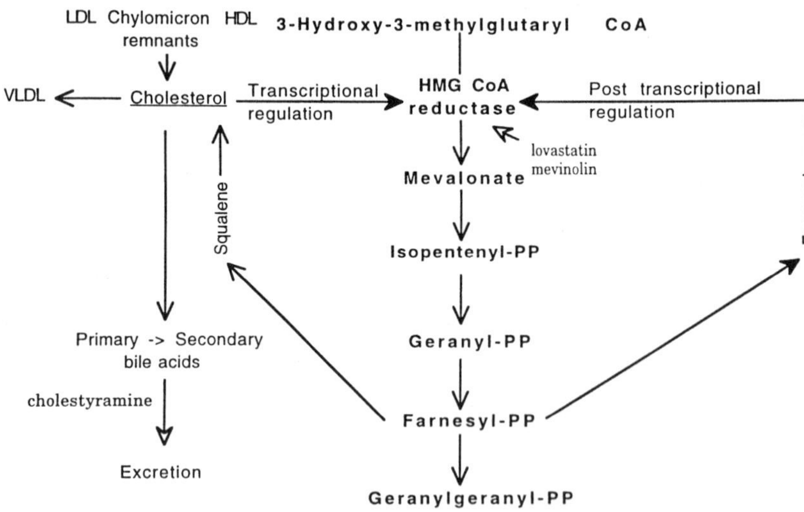

Figure 2. Intermediate and end products of mevalonate metabolism in animal tissues. HMG CoA reductase is the rate-limiting activity for the synthesis of mevalonic acid, the rate-limiting substrate for the synthesis of the cholesterol.[7] Cholesterol and farnesol, respectively, act as the signals for the transcriptional and post transcriptional regulation of HMG CoA reductase activity insterologenic tissues.[7, 12] Competitive inhibitors of HMG CoA reductase activity (Lovastatin and related products) are widely prescribed for the control of hypercholesterolemia.[7] HMG CoA reductase activity in tumor cells is elevated and resistant to sterol feedback regulation.[15-20]

Mevalonate pathway activities are present in and essential to the survival of all plant and animal tissues. Elements of the pathway also offer targets for regulating plant growth and senescence and for the control of insects, bacteria and fungus.

2. 3-HYDROXY-3-METHYLGLUTARYL COENEZYME A REDUCTASE

Goldstein and Brown[7] provide a comprehensive review of the fine tuning of 3-hydroxy-3-methylglutaryl coenzyme A (HMG CoA) reductase, the product of which is the precursor of the diverse isoprenoid groups that are incorporated into the end products of the mevalonate pathway. The sterol-mediated regulation of transcription is the major effector of this and other mevalonate pathway activities in animal cells (Figure 3). As a consequence, cholesterol delivered to the liver via chylomicron remnants, HDL and LDL and to peripheral tissues, predominantly by LDL, down regulates cholesterol synthesis. In tissues deprived of cholesterol by treatment with a low cholesterol diet, soluble fiber, cholestyramine or lovastatin there is a compensatory increase in HMG CoA reductase mRNA and mass and associated mevalonate pathway activities.

A second level of control (post transcriptional regulation, Figure 3) permits the continuation of mevalonate synthesis in the presence of saturating concentrations of cholesterol.[7] Then when the mevalonate requirement for the high affinity synthesis of non-sterol products is satisified, a mevalonate-derived, non-sterol signal triggers the non-lysosomal

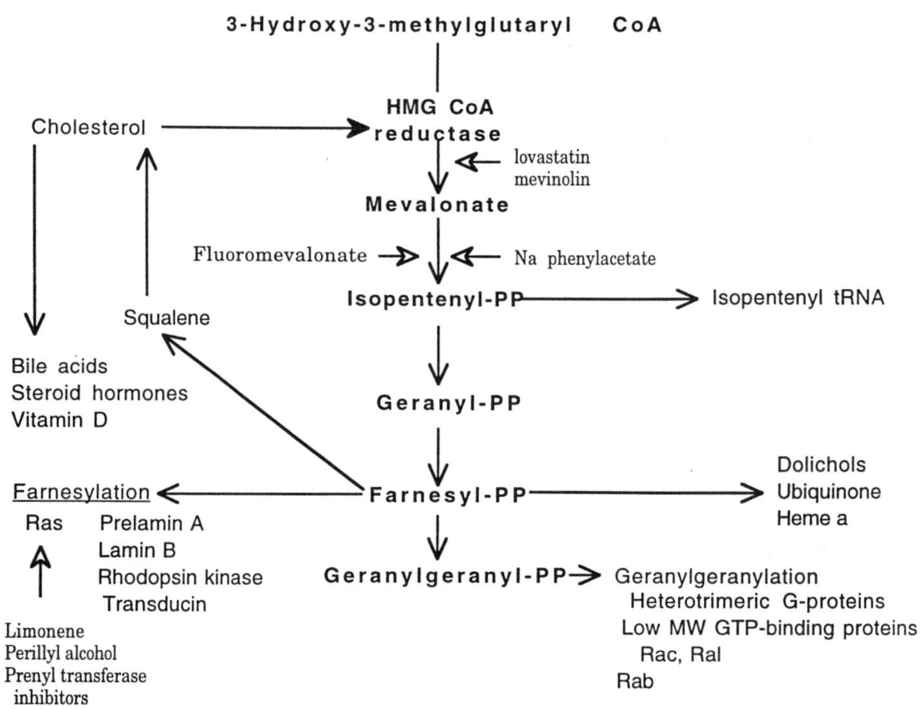

Figure 3. End products of mevalonate metabolism that play essential roles in the survival of animal tissues. The suppressors of mevalonate pathway activities (Lovastatin,[7] Sodium phenylacetate[8], Fluoromevalonate[9], Limonene/Perillyl alcohol[64] and Prenyl transferase inhibitors[7] have potential as chemotherapeutic agents.

degradation of HMG CoA reductase and secondarily, decreases the rate of translation of HMG CoA mRNA. HMG CoA reductase is reversibly inactivated by phosphorylation catalyzed by an AMP-activated protein kinase.[7] Phosporylation of HMG CoA reductase also accelerates its degradation.[10] The tentative identification of farnesol (Figure 4) as the mevalonate-derived, non-sterol signal[11, 12] draws support with the recent demonstration of a nuclear farnesol receptor that has low affinity for farnesal and farnesoic acid.[13] In plant tissues, the non-sterol modulator of HMG CoA reductase degradation and a kinase appear to play the major roles in regulating mevalonate synthesis.[14] Restated, mevalonate synthesis in sterologenic animal tissues is regulated principally at the level of transcription and in non-sterologenic plant, insect and microbial cells, by post transcriptional events.

Central to our hypothesis are the findings of the elevated HMG CoA reductase activity and its uncoupling from sterol-mediated feedback regulation in tumors and embryonic and differentiating tissues and in carcinogen-treated and regenerating liver.[15-20] The presumption

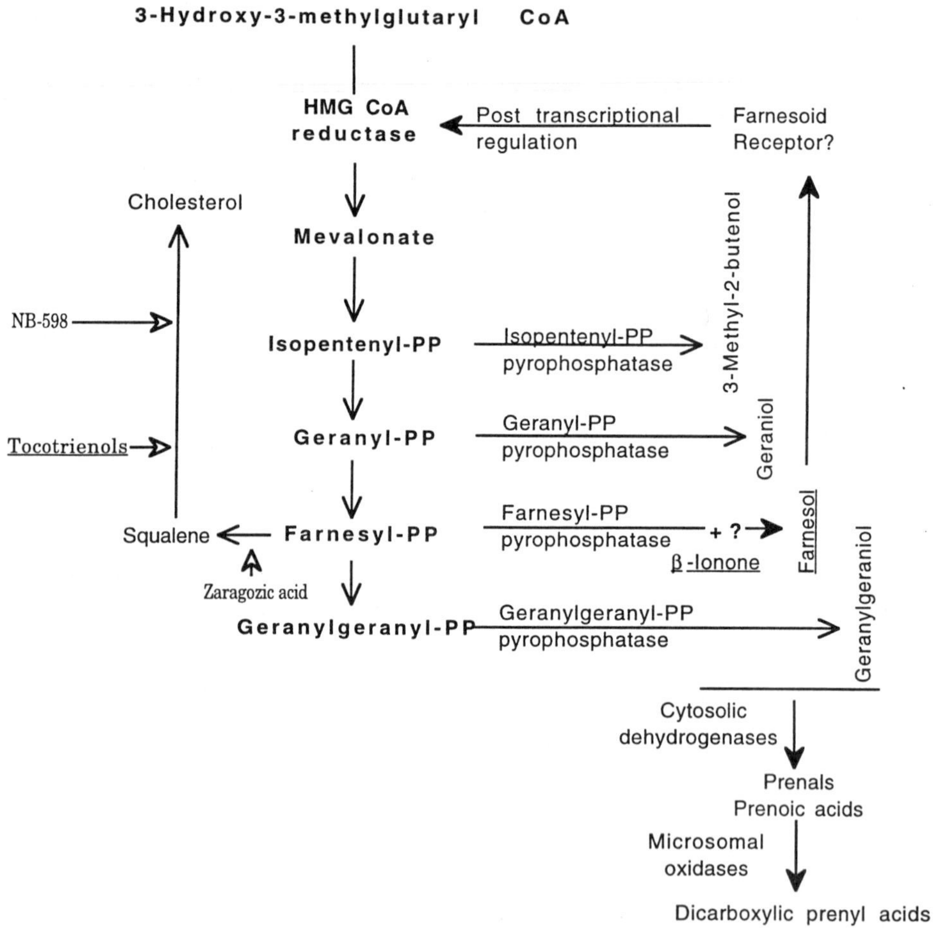

Figure 4. Tocotrienols,[38] NB-598 and zaragozic acid[11, 12] inhibit mevalonate pathway activities at the indicated sites; each is postulated to elevate cellular farnesyl-PP. Based on our finding that β-ionone induces geranyl-PP pyrophosphatase we postulate that it similarly induces farnesyl-PP pyrophosphatase activity.[43] If so, β-ionone is postulated to potentiate the impact of the tocotrienols, NB-598 and zaragozic acid on the post transcriptional regulation of HMG CoA reductase activity. Microsomal oxidase activities induced by pentobarbital attenuate the impact of β-ionone on HMG CoA reductase activity.[32]

that cholesterol, the bulk end-product of mevalonate metabolism and an important membrane constituent, was essential for growth ended with the finding that competitive inhibitors of mevalonate synthesis arrested cell division at the G1/S interface of the cell cycle[21-23]; growth could be restored with mevalonate but not with cholesterol supplements.[22-25]

Among the regulation-related events occurring prior to the onset of the S phase is the enhanced synthesis of HMG CoA reductase[23] which is resistant to sterol feedback regulation.[26] Asynchronous[25] and synchronous[21-23] cells arrested at the G1/S interface of the cell cycle exhibit an altered morphology. The report[27] of the posttranslational incorporation of farnesyl-pyrophosphate and geranylgeranyl-pyrophosphate into nuclear and soluble proteins provided the first evidence of a non-steroidal role for mevalonate (Figure 2). A broad spectrum of cellular proteins undergoes isoprenylation,[28, 29] the specific proteins which are isoprenylated depends upon the availability of mevalonate.[30]

2.1. Dietary Isoprenoids Suppress Hepatic HMG CoA Reductase Activity

Our anomalous findings that, contrary to the action of bile acid sequestrants, the fiber-rich grains [31] and alfalfa[32] suppressed hepatic HMG-CoA reductase activity led to the recognition that minor constituents of barley and alfalfa, α-tocotrienol and β-ionone, elicit *in vivo*, dose-dependent suppressions of reductase activity. β-Ionone is both a natural constituent and a product of carotenoid degradation (Figure 1). Monoterpenoid and sesquiterpenoid products of plant mevalonate pathways (Figure 1) also suppress hepatic HMG-CoA reductase activity.[33-40] Unlike the competitive inhibitors of HMG-CoA reductase activity, the inhibitory action of the tocotrienols[38,39] and the cyclic monoterpenes[33] on HMG-CoA reductase activity is mediated via a decrease in enzyme mass. The post-transcriptional action is postulated as γ-tocotrienol,[38] β-carotene,[39] farnesyl acetate,[40] farnesyl-pyrophosphate and farnesol[11, 12] do not lower HMG CoA reductase mRNA levels whereas both reductase mass and activity are substantially decreased. The inhibitory action is time- and concentration-dependent. For example, Parker et al.[38] report the 20% and 80% suppression of HMG CoA reductase activity in HepG2 cells with 1 and 10 μM γ-tocotrienol. ^{35}S Labelled HMG CoA reductase in HepG2 cells in 10 μM γ-tocotrienol decreased linearly during a 5 hr incubation whereas other ^{35}S labelled proteins remained intact. Incorporation of acetate, but not of mevalonate, into sterols parallels the decline in HMG CoA reductase and reductase activity. Parker et al.[38] concluded that γ-tocotrienol specifically modulates the intracellular mechanism for the controlled degradation of the reductase protein. This activity mirrors post transcriptional action of farnesol[11, 12] a diverted product of the mevalonate pathway[41] which is now recognized to be the physiological, non-sterol isoprenoid regulator of HMG CoA reductase activity.[7] The prediction that the prenyl alcohol rather than the prenyl acid acts as the regulatory molecule[41] draws support from the recent detection of a nuclear receptor for farnesol which has low affinity for farnesoic acid.[13] Cytosolic prenyl alcohol dehydrogenase and microsomal oxidase activities (Figure 4) catalyze the conversion of farnesol and other prenyl alcohols to prenoic acids and dicarboxylic prenoic acids prior to their excretion.[42] β-Ionone suppresses hepatic HMG CoA reductase activity with a concomitant lowering of serum cholesterol.[32] We further examined the impact of dietary β-ionone on hepatic prenoic acid synthesis. We found that this treatment induced microsomal geranyl-pyrophosphate pyrophosphatase activity but had no impact on cytosolic prenyl alcohol dehydrogenase activities.[43] These findings suggest that dietary mono- and sesquiterpenes may induce farnesyl-pyrophosphate pyrophosphatase activity.[44] The tocotrienols, acting later in the mevalonate pathway,[38] are postulated to increase the cellular pool of farnesyl-pyrophosphate. These actions we suggest result in an elevated level of cellular farnesol leading to the post transcriptional down regulation of HMG CoA reductase activity (Figures 4).

2.2. Relating the Suppression of HMG CoA Reductase Activity to the Cholesterol Lowering Action of Dietary Isoprenoids

A 5% suppression of hepatic HMG CoA reductase activity by dietary isoprenoids is associated with a 2% lowering of serum cholesterol ($r^2=0.62$, $P<0.004$).[43] This observation is offered with the caveat that it applies only to animal studies employing cholesterol-free diets. The impact of dietary isoprenoids on human cholesterol levels are less certain. Geraniol[36] and a natural blend of tocotrienols (Palmvitee)[45-47] had either a modest[36, 45, 46] or no[47] impact on cholesterol levels of hypercholesterolemic human subjects. A subset of subjects within each experimental population responded to the treatment. Interpretation of these studies is further complicated by the genetic diverstiy of the subjects and by differences in the composition of the natural blends of tocotrienols. α-Tocopherol suppresses the cholesterol-suppressive efficacy of dietary tocotrienols (Qureshi, Pearce, Nor, Gapor, Peterson, & Elson, submitted). Under a controlled dietary regimen (AHA Step I diet) 200 mg supplements of low tocopherol- (Qureshi, Bradlow, Brace, Manganello, Peterson, Pearce, Wright, Gapor & Elson, submitted) and tocopherol-free (Qureshi, Bradlow, Salser, Brace, Manganello, & Elson, submitted) tocotrienol preparations elicited 10-13% decreases in the cholesterol levels of hypercholesterolemic subjects.

2.3. Isoprenoids Suppress the Growth of Chemically-Initiated and Transplanted Tumors and Cause the Regression of Established Tumors

Dietary isoprenoids suppress carcinogenesis initiated by direct acting carcinogens and by carcinogens requiring metabolic activation, suppress the growth of transplanted tumors, and cause the regression of established, chemically-initiated tumors.[1-4] The studies reviewed here are limited to those of the antitumor actions of β-ionone and the tocotrienols. Dietary β-ionone increased ($P<0.001$) the latency of tumor formation in the DMBA-initiated rat mammary tumor model and reduced tumor burden by 50%.[48] A high-tocopherol Palmvitee preparation increased ($P<0.03$) the median latency of tumor development in the same model.[49]

Other studies showed that the tocol blend delayed the appearance of DMBA-initiated, croton oil-promoted papillomas and the formation of virally induced lymphoma in genetically recessive HTS/J female mice.[50] The blend also reduced the severity of 2-acetylaminofluorene-induced hepato-carcinogenesis.[51] Massive doses of the tocols were administered in these studies. Recent findings suggest that a much lower dose of a single tocotrienol might be more effective than the tocol blend. Consistent with our findings noted above, (Qureshi, Pearce, Nor, Gapor, Peterson, & Elson, submitted) the IC_{50} for the inhibition of sterol synthesis by the blend is more than double that for γ- or δ-tocotrienol.[37]

A chemotherapeutic action of purified tocotrienols is suggested by studies showing that the survival times of mice bearing sarcoma 180, Ehrlich carcinomas and IMC carcinomas are significantly increased by injections of α- or γ-tocotrienol whereas injections of α-tocopherol were ineffective.[52]

In vitro assays confirm the broad tumor-suppressive actions of the tocotrienols[53-55], β-ionone and geraniol (Table 1). The related compounds, β-carotene nor α-tocopherol exhibited little if any antitumor activity. Tumor types differed substantially in sensitivity to the isoprenoids. The expression of Epstein-Barr virus early antigen expression in a Raji cell line (Epstein-Barr virus-genome-carrying human lymphoblastoid cells) is inhibited by 50 μM γ- and δ-tocotrienol.[56]

Table 1. IC50 values[1] for the suppression of tumor cell growth by geraniol, b-ionone and the tocotrienols

	Isoprenoid	μM	Reference
P388 mouse lymphoid	α-tocopherol	2.320	52
P388 mouse lymphoid	α-tocotrienol	0.100	52
P388 mouse lymphoid	γ-tocotrienol	0.050	52
P388 mouse lymphoid	geraniol	0.500	57
B16 mouse melanoma	geraniol	0.100	57
B16 mouse melanoma	α-tocopherol	>1.000	Elson[2]
B16 mouse melanoma	α-tocotrienol	0.110	Elson[2]
B16 mouse melanoma	γ-tocotrienol	0.015	Elson[2]
B16 mouse melanoma	δ-tocotrienol	0.009	Elson[2]
B16 mouse melanoma	didesmethyl tocotrienol	0.0008	Elson[2]
B16 mouse melanoma	β-carotene	>1.000	Elson[2]
B16 mouse melanoma	β-ionone	0.120	Elson[2]
Rat mammary epithelial	β-ionone	>1.000	Sheffield[2]
H69 small cell lung carcinoma	α-tocopherol	2.320	52
H69 small cell lung carcinoma	α-tocotrienol	0.100	52
H69 small cell lung carcinoma	γ-tocotrienol	0.100	52
HeLa cervix carcinoma	α-tocopherol	2.320	52
HeLa cervix carcinoma	α-tocotrienol	0.100	52
HeLa cervix carcinoma	γ-tocotrienol	0.100	52
MDA-MB-435 breast cancer	α-tocopherol	>0.418	55
MDA-MB-435 breast cancer	δ-tocotrienol	0.220	55
MDA-MB-435 breast cancer	α-tocotrienol	0.220	55
MDA-MB-435 breast cancer	γ-tocotrienol	0.070	55
MCF-7 breast cancer	β-ionone	0.010	Sheffield[2]
MDA-MB-231 breast cancer	β-ionone	0.100	Sheffield[2]

[1] concentration of an isoprenoid required to suppress net increase in cell count by 50%
[2] University of Wisconsin, unreported

2.4. Isoprenoids Partially Arrest Asynchronous Cells at the G1/S Interface of the Cell Cycle

Shoff et al.[57] reported the incomplete arrest of B16 melanoma cells cultured in 50-150 μM geraniol. The profile obtained with flow cytometric analysis matched that of asynchronous cells arrested with a competitive inhibitor of mevalonate synthesis.[25]

2.5. Farnesylation: Isoprenoids Inhibit the Farnesylation of Small G-Proteins

Our observation that an isoprenoid, d-limonene, and its more potent serum metabolites, inhibit protein isoprenylation with some specificity for the 21-26 kDa proteins[58] has been confirmed.[59-63] Crowell's group[58, 62] incubated NIH3T3 in media containing 30 μM lovastatin and [2-^{14}C] mevalonolactone in the absence or presence of diverse monoterpenes for 3 hr; 1 mM geraniol and derivatives of perillic acid, a hepatic metabolite of d-limonene inhibited the farnesylation (incorporation of [2-^{14}C] mevalonolactone) of 21-26 kDa proteins by 50%. More recently Hohl and Lewis[63] report that monoterepenes either depress the synthesis of ras or increase the rate of ras degradation. The apparent specificity for the inhibition of ras synthesis (or increase in ras degradation) and isoprenylation suggested to some investigators that d-limonene and perillyl alcohol would be good candidates for clinical evaluation.[64] However some investigators[65, 66] report that the monoterpene action involves factors other than ras (reviewed in reference[1]).

2.6. Farnesylation: the Inhibition of the Farnesylation of ras does not Explain the Tumor-Suppressive Action of Lovastatin and Isoprenoids

NIH 3T3 cells transformed by c-ras and v-ras oncogenes (ras farnesylation-dependent) and by v-src, v-raf and myristolated ras oncogenes (ras farnesylation-independent) were found to be equally sensitive to growth inhibition by lovastatin. Lovastatin blocked the processing of endogenous ras and v-ras proteins yet had no effect on the lipidation of the myristolated ras. Supplemental mevalonate reversed the effect of lovastatin.[65] DeClue et al.[66] and Ruch and Sigler[65] conclude that the growth-suppressive action of lovastatin and limonene is likely due to the inhibition of the processing of other cellular proteins. Early investigations[25, 27, 67] demonstrated that lovastatin blocked the incorporation of acetate into nuclear proteins.

3. APOPTOSIS

Classic apoptosis (programmed cell death) is a genetically encoded cell death program, defined by characteristic morphologic and biochemical changes, which progresses under tightly controlled circumstances.[68-70] A principal biochemical indicator of apoptosis is the internucleosomal fragmentation of DNA, an effect attributed to the induction or activation of an endogenous nuclease, and identified by the detection of DNA "ladders" (50 and 300 kb fragments) following agarose gel electrophoresis.[69] Cytotoxic drugs at low doses may initiate apoptotic-like responses whereas higher doses cause necrosis. The apoptotic-like response triggered by drugs has many characteristics of classic apoptotic cell death but may lack the appearance of 300 kb DNA fragments. Apoptosis can also be detected by flow cytometric analysis. A key feature of the DNA frequency distribution histogram of apoptotic cells is the presence of a sub-G1 peak.[71]

3.1. Lovastatin Induces Apoptosis

At low concentrations (≤ 10 μM) lovastatin suppresses cell growth by >50%; the inhibition is reversed by supplemental mevalonate; at higher concentrations (≥ 25 μM) lovastatin has a clear cytotoxic effect.[57, 72] The morphological and DNA characteristics of viable (trypan blue-excluding) human promyelocytic HL-60 cells after incubation in 10 μM lovastatin are compatible with many characteristics of apoptotic cell death.[72]

3.2. Do Isoprenoids Induce Apoptosis?

The proliferation of tumor cells is suppressed by α-, δ- and γ-tocotrienol, β-ionone and geraniol (Table 1). Melanoma cells incubated with geraniol are partially arrested at the G1/S interface of the cell cycle.[57] Addition of an isoprenoid to a 50% confluent culture of B16 cells causes a fraction of the cells to round up and detach from the monolayer much as occurs when cells are treated with lovastatin. A preliminary assessment of the cells remaining on the monolayer showed that, when washed and reincubated in fresh medium, trypan blue-excluding cells from ≤ 80 μM β-ionone cultures form colonies whereas those from ≥ 160 μM β-ionone cultures remain in suspension and gradually disappear. The DNA histograms of melanoma[57] and leukemia[73] cells arrested with geraniol and farnesol show a prominent sub-G1 peak, a feature of apoptopic cells.[71]

Following the reports that farnesol acts as the intracellular, post transcriptional modulator of HMG CoA reductase activity[9, 10] we made a preliminary assessment of farnesol levels in B16 cells following exposure to γ-tocotrienol. The concentration of a metabolite which co-eluted with a farnesol standard was markedly elevated in cells incubated with greater than IC50 concentrations (15 μM) of γ-tocotrienol. We have not confirmed by mass spectral analysis the identity of the co-eluting metabolite. While this preliminary observation awaits confirmation, the results suggest that isoprenoids induce an apoptotic-like response via an increase in cellular farnesol.

3.2.1. the Potency with Which an Isoprenoid Suppresses HMG CoA Reductase Activity Predicts Its Potency in Suppressing the Growth of Tumors and the Proliferation of Tumor Cells. This hypothesis is explored in recent reviews.[2,3] We propose that tumor growth is limited by the size of the mevalonate pool supporting the synthesis of the prenyl-pyrophosphates. As a consequence, the HMG CoA reductase activity in a culture increases in parallel with the increase in cell population; thus, HMG CoA reductase specific activity remains constant across a range in the rate of cell proliferation. We avoid this pitfall in showing that the potency with which an isoprenoid suppresses hepatic HMG CoA reductase specific activity predicts with very high accuracy, the potency of the isoprenoid in tests of the post-initiation suppression of tumor growth. Plots exploring the relationship between impact of tocols,[37] mono- and sesquiterpenes[32,34,74] on hepatic HMG CoA reductase and the impact of these isoprenoids on tumor-bearing animals [respectively 52; 48, 75] yield r^2 values approaching unity.[2] Although obvious limits apply to the biological interpretation of these plots, the mean r^2, 0.90±0.05, clearly shows that the potency with which an isoprenoid suppresses hepatic HMG CoA reductase activity accurately predicts that isoprenoid's anti tumor action.

3.2.2. Lovastatin, Pravastatin, Simvastatin and Mevinolin Suppress Tumorigenesis; Cholestyramine and Carrageenan Promote Tumorigenesis. The hypocholesterolemic drugs that inhibit HMG Co A reductase activity suppress the growth of transplanted neuroblastoma,[76] pancreatic carcinoma,[77] melanoma[78] and ascites hepatomas.[79] Lovastatin is in clinical trial for the treatment of malignant gliomas.[80,81] These inhibitors also suppress 1,2-dimethylhydrazine-induced colon tumorigenesis.[82] The levels of the drugs used in these experiments are more than 5-times higher than those used in the treatment of human hypercholesterolemia. Never the less findings from a 5-year safety and efficacy study of the management of hypercholesterolemia point to a lovastatin-mediated 50% reduction in cancer incidence.[83] The second approach to treating hypercholesterolemia, the bile acid sequestrants, induces HMG CoA reductase activity.[7] As might be anticipated, cholestyramine and carrageenan promote, respectively, 7,12-dimethylbenzanthracene-induced mammary tumorigenesis[84] and 1,2-dimethylhydrazine-induced colon tumorigenesis.[85]

4. TARGETING THE ACTION OF THE ISOPRENOIDS TO TUMORS

Diet/cancer and diet/cardiovascular disease interrelationships may be explained by the mevalonate-suppressive action of isoprenoid end products of plant secondary metabolism. Assorted monoterpenes, sesquiterpenes, carotenoids and tocotrienols post-transcriptionally down regulate HMG CoA reductase activity, a key activity in the sterologenic pathway. The modest decrease in cholesterol synthesis is associated with a concomitant lowering of LDL cholesterol. Bennis et al.[20] confirm previous reports that HMG CoA reductase activity in tumor cells is elevated and resistant to sterol feedback regulation.[15-19] Their data show that 40% (160 pmol/min/mg protein) of the HMG CoA reductase activity in human lung adenocarcinoma cells is devoted to the synthesis of non-sterol products whereas non-sterol isoprenoid synthesis in normal fibroblasts accounts for only 15-20% (~40 pmol/min/mg protein) of the activity. As pointed out in our reviews[1-4] mevalonate synthesis in neoplastic cells tends to be resistant to sterol feedback regulation while retaining sensitivity to post-transcriptional control. A rationale for the elevated and sterol feedback resistant HMG CoA reductase activity in tumors appears in recent reports. The HMG CoA reductase gene in hepatic nodules generated during chemical carcinogenesis[86, 87] and in colon polyps and cancers[88] is hypomethylated and over expressed. 1,2-Dimethylhydrazine and aristolochic acid respectively interact (predominantly) with the guanine and adenine residues of DNA. Never the less, similar patterns of hypomethylation patterns are generated by the two carcinogens. Sarma's group[86, 87] is constructing probes to determine the exact locations of the hypomethylated sequences in the hepatic gene. These important studies may explain both the over expression and the sterol-feedback resistance of the tumor enzyme. The tumor enzyme on the other hand is extremely sensitive to isoprenoid-mediated down regulation. For example, mevalonate synthesis in HepG2 cells (IC50 = 1.6 μM[38]) is several times more sensitive than that in primary hepatocytes (IC50 = 500 μM[37]) to down regulation by γ-tocotrienol. The regulatory scheme proposed by Goldstein and Brown[7] might explain the unique sensitivity of the tumor cell to dietary isoprenoids. Maximum post transcriptional down regulation of reductase activity requires both cholesterol and the endogenous, mevalonate-derived regulator (farnesol[11, 12]).

5. DIET

β-Ionone, an end-ring analog of β-carotene, and its oxygenated analogs are widely distributed in plant products in free and conjugated forms.[89-109] Tocotrienols are present in cereal grains, principally oats, barley and rye,[110-114] and in oils extracted from rice bran, barley, palm fruit and olives.[115-119] Genotype and environmental factors play major roles in determining the profile of the isoprenoids present in a given species.[110] Differences in storage conditions and processing technologies further influence the concentration of a specific isoprenoid in a plant product.

The above considerations may explain the conflicting reports of the impact of oats, bran, alcohol and milk on serum cholesterol levels. Some researchers record significant impacts whereas others report finding no effect of oat products on cholesterol levels (reviewed in [120]). There are also conflicts regarding the mechanism of the action of oats products.[120-122] Anderson's laboratory[120] postulates that volatile fatty acids produced by the fermentation of soluble oat fiber suppress cholesterol synthesis whereas other laboratories[121-123] point to a tocotrienol-mediated action, an action consistent with findings previously discussed.[31, 37, 38] The findings that cereal grains vary widely in their tocotrienol content[110] reconciles the conflicting reports of the cholesterol-lowering efficacy of oats products providing that the responsible agent is a tocotrienol rather than a volatile fatty acid.

An analogous situation may apply to brans. Rice bran,[124, 125] with an extremely low β-glucan content, is as effective as high β-glucan oat bran in lowering serum cholesterol.[126] The lowering of cholesterol levels by rice bran,[125, 126] like that of oat bran,[125, 127, 128] may be in part attributed to cholesterol-suppressive actions of the tocotrienols.

β-Ionone imparts a woody, violet off-flavor to milk.[129, 130] Some researchers attribute its presence to the oxidative degradation of β-carotene and retinyl palmitate.[129] On the other hand, our study suggests that β-ionone is a passive constituent whose presence in milk reflects the composition of the dairy ration.[32] Passive constituents we suggest are responsible for the cholesterol-suppressive action sometimes reported for milk. β-Ionone gives a desirable fruity flavor to wine.[100, 101] The French paradox suggests that non-ethanol constituents of wine have benefits for cardiovascular disease independent of ethanol effects.[131] β-Ionone is a likely contributor of this benefit.

There are no data bases for estimating dietary intakes of diverse isoprenoids. Likely additive effects of isoprenoids have not been delineated. Human studies are limited. Never the less, there are sufficient data to support a conclusion that a supplemental intake of 0.5 to 1 mmol of an isoprenoid is sufficient to have an impact on human cholesterol metabolism[36, 45, 46] (Qureshi, Bradlow, Brace, Manganello, Peterson, Pearce, Wright, Gapor & Elson, submitted, (Qureshi, Bradlow, Salser, Brace, Manganello, & Elson, submitted).

REFERENCES

1. Elson, C.E. & Yu, S.G. (1994) The chemoprevention of cancer by mevalonate-derived constituents of fruits and vegetables. J. Nutr. 124: 607-614.
2. Elson, C.E. (1995) Suppression of mevalonate pathway activities by dietary isoprenoids: Protective roles in cancer and cardiovascular disease. J. Nutr. 125: 1666s-1672s
3. Elson, C.E. & Qureshi, A.A. (1995) Coupling the cholesterol- and tumor-suppressive actions of palm oil to the impact of its minor constituents on 3-hydroxy-3-methylglutaryl coenzyme A reductase activity. Prostaglandins Leukotrienes Essent. Fatty Acids: 52:205-208
4. Elson, C.E., Yu, S.G. & Qureshi, A.A. The cholesterol- and tumor-suppressive actions of palm oil isoprenoids. Lipids: (in press)
5. Block, G., Patterson, B. & Subar, A., Fruit, vegetables, and cancer prevention: A review of the epidemiological evidence. Nutr. Cancer, 18:1-29, 1992.
6. Heinonen, O.P. & Albanes, D. (The Alpha-Tocopherol, Beta Carotene Cancer Prevention Study Group) The effects of vitamin E and beta carotene on the incidence of lung cancer and other cancers in male smokers. NEJM, 330:1029-1035, 1994.
7. Goldstein, J.S. & Brown, M.S. (1990) Regulation of the mevalonate pathway. Nature (London) 343: 425-430.
8. Ram, Z., Samid, D., Walbridge, S., Oshiro, E.M., Viola, J.J., Tao-Cheng, J.H., Shack, S., Thibault, A., Myers, C.E. & Oldfield, E.H. (1994) Growth inhibition, tumor maturation, and extended survival in experimental brain tumors in rats treated with phenylacetate. Cancer Res. 54:2923-7.
9. Cuthbert, J.A. & Lipsky, P.E. (1995) Suppression of the proliferation of Ras-transformed cells by fluoromevalonate, an inhibitor of mevalonate metabolism. Cancer Res. 55:1732-40.
10. Zammit, V.A. & Easom, R.A. (1987) Regulation of hepatic HMG-CoA reductase *in vivo* by reversible phosphorylation. Biochim. Biophys. Acta 927:223-228.
11. Correll, G.C. & Edwards, P.A. (1994) Mevalonic acid-dependent degradation of 3-hydroxy-3-methylglutaryl-coenzyme A reductase *in vivo* and *in vitro*. J. Biol. Chem., 269:633-638.
12. Correll, C.C., Ng, L. & Edwards, P.A. (1994). Identification of farnesol as the non-sterol derivative of mevalonic acid required for the accelerated degradation of 3-hydroxy-3-methylglutaryl coenzyme A reductase. J. Biol. Chem. 269:17390-17393.
13. Forman, B.M., Goode, E., Chen, J., Oro, A.E., Bradley, D.J., Perlmann, T., Noonan, D.J., Burka, L.T., Mc Morris, T., Lamph, W.W., Evans, R.M. & Weinberger, C., Identification of a nuclear receptor that is activated by farnesol metabolites. Cell, 81:687-693, 1995.
14. Stermer, B.A., Bianchini, G.M. & Korth, K.L. (1994) Review: Regulation of HMG-CoA reductase activity in plants. J. Lipid Res. 35:1133-1140.
15. Siperstein, M.D. & Fagan, V.M. (1964) Deletion of the cholesterol-negative feedback system in liver tumors. Cancer Res. 24:1108-1115.
16. Yachnin, S., Toub, D.B. & Mannickarottu, V. (1984) Divergence in cholesterol biosynthetic rates and 3-hydroxy-3-methylglutaryl-CoA reductase activity as a consequence of granulocyte versus monocyte-macrophage differentiation in HL-60 cells. Proc. Natl. Acad. Sci. USA 81:894-897.
17. Bruscalalupi, G., Leoni, S., Mangiantini, M.T., Minieri, M., Spagnuolo, S. & Trentlanc, A. (1985) True uncoupling between cholesterol synthesis and 3-hydroxy-3-methylglutaryl coenzyme A reductase in an early stage of liver generation. Cell. Molec. Biol. 31:365-368.
18. Engstrom, W. & Schofield, P.N. (1987) Expression of the 3-hydroxy-3-methyl-glutaryl coenzyme A-reductase and LDL-receptor genes in human embryonic tumors and in normal fetal tissues. Anticancer Res. 7:337-342.
19. Azrolan, N.I. & Coleman, P.S. (1989) A discoordinant increase in the cellular amount of 3-hydroxy-3-methylglutaryl-CoA reductase results in the loss of rate-limiting control over cholesterogenesis in a tumor cell-free system. Biochem. J. 258:421-425.

20. Bennis, F., Favre, G., Le Gaillard, F. & Soula, G. (1993) Importance of mevalonate-derived products in the control of HMG-CoA reductase activity in the growth of human lung adenocarcinoma cell line A549. Int. J. Cancer 55:640-645.
21. Sinenski, M. & Logel, J. (1985) Defective macromolecule biosynthesis and cell cycle progression in a mammalian cell starved for mevalonate. Proc. Natl. Acad. Sci. USA 82:3257-3261.
22. Langan, T.J. & Volpe, J.J. (1987) Cell cycle-specific requirement for mevalonate but not for cholesterol, for DNA synthesis in glial primary cultures. J. Neurochem. 49:513-521.
23. Doyle, J.W. & Kandutsche, A.A. (1988) Requirement for mevalonate in cycling cells: quantitative and temporal aspects. J. Cell. Physiol. 137:133-140.
24. Fairbanks, K.D., Barbu, V.D., Witte, L.D., Weinstein, I.B. & Goodman, D.S. (1986) Effects of mevinolin and mevalonate on cell growth in several transformed cell lines. J. Cell Physiol. 127:216-222.
25. Maltese, W.A. & Sheridan, K.M. (1987) Isoprenylated proteins in cultured cells: Subcellular distribution and changes related to altered morphology and growth arrest induced by mevalonate deprivation. J. Cell. Physiol. 133:471-481.
26. Panini, S.R., Schnitzer-Polokoff, R., Spencer, T.A. & Sinenski, M. (1989) Sterol-independent regulation of 3-hydroxy-3-methylglutaryl-CoA reductase by mevalonate in Chinese hamster ovary cells. J. Biol. Chem. 264:11044-11052.
27. Schmidt, R.A., Schneider, C.J. & Glomset, J.A. (1984) Evidence for post-translational incorporation of a product of mevalonic acid into Swiss 3T3 cell proteins. J. Biol. Chem. 259:10175-10180.
28. Hancock, J.F., Magee, A.I., Childs, J.E. & Marshall, C.J. (1989) All ras proteins are polyisoprenylated but only some are palmitoylated. Cell 57:1167-1177.
29. Schafer, W.R., Kim, R., Sterne, R., Thorner, J., Kim, S-H. & Rine, J. (1989) Genetic and pharmacological suppression of oncogenic mutations in ras genes of yeast and humans. Science 245:379-385.
30. Rilling, H.C., Bruenger, E., Leining, L.M., Buss, J.E. & Epstein, W.W. (1993) Differential prenylation of proteins as a function of mevalonate concentration in CHO cells. Arch. Biochem. Biophys. 301:210-215.
31. Qureshi, A.A., Burger, W.C., Peterson, D.A. & Elson, C.E. (1986) The structure of an inhibitor of cholesterol biosynthesis isolated from barley. J. Biol. Chem. 261:10544-10550.
32. Yu, S.G., Abuirmeileh, N.M., Qureshi, A.A. & Elson, C.E. (1994). Dietary β-ionone suppresses hepatic 3-hydroxy-3-methylglutaryl coenzyme A reductase activity. J. Agric. Food Chem. 42:1493-1496.
33. Clegg, R.J., Middleton, B., Bell, G.D. & White, D. A. (1982) The mechanism of cyclic monoterpene inhibition of 3-hydroxy-3-methylglutaryl coenzyme A reductase *in vivo* in the rat. J. Biol. Chem. 257:2294-2299.
34. Qureshi, A.A., Mangels, A.R., Din, Z.Z. & Elson, C.E. (1988) Inhibition of hepatic mevalonate biosynthesis by the monoterpene, d-limonene. J. Agric. Food Chem. 36:1220-1224.
35. Fitch, M.E., Mangels, A.R., Altmann, W.A., El Hawry, M., Qureshi, A.A. & Elson, C.E. (1989) Microbiological screening of mevalonate-suppressive minor plant constituents. J. Agric. Food Chem. 37:687-691.
36. Elson, C.E., Underbakke, G.L., Hanson, P., Shrago, E., Wainberg, R. & Qureshi, A. (1989) Impact of lemongrass oil, an essential oil, on serum cholesterol. Lipids 24:677-679.
37. Pearce, B.C., Parker, R.A., Deason, M.E., Qureshi, A.A.& Wright, J.J.K. (1992) Hypocholesterolemic activity of synthetic and natural tocotrienols. J. Med. Chem. 35:3595-3606.
38. Parker, R.A., Pearce, B.C., Clark, R.W., Gordan, D.A. & Wright, J.J.K. (1993) Tocotrienols regulate cholesterol production in mammalian cells by post-transcriptional suppression of 3-hydroxy-3-methylglutaryl-coenzyme A reductase. J. Biol. Chem. 268, 11230-11238.
39. Moreno, F., Rossiello, M.R., Manjeshwar, S., Nath, R., Rao, P.M., Rajalakshmi, S. & Sama, D.S.R. (1994) Effect of β-carotene on the expression of 3-hydroxy-3-methylglutaryl coenzyme A reductase in rat liver. Proc. AACR 35: 142 (abs.).
40. Bradfute, D.L. & Simoni, R.D. (1994) Non-sterol compounds that regulate cholesterogenesis. Analogues of farnesyl pyrophosphate reduce 3-hydroxy-3-methylglutaryl-coenzyme A reductase levels. J. Biol. Chem. 269: 6645-6650.
41. Gonzales-Pacanowska, D., Arisan, B., Havel, C.M. & Watson, J.A. (1988) Isopentenoid synthesis in isolated embryonic Drosophilacells. Farnesl catabolism and omega oxidation. J. Biol. Chem. 263:1301-1306.
42. Christophe, J. & Popják, G. (1961) Studies on the biosynthesis of cholesterol: XIV. the origin of prenoic acids from allyl pyrophosphates in liver enzyme systems. J. Lipid Res. 2:244-257, 1961.
43. Case, G.L., He, L., Mo, H. & Elson, C.E. Induction of geranyl-pyrophosphate pyrophosphatase activity by cholesterol-suppressive isoprenoids. Lipids 30:357-359, 1995.
44. Bansal, V.S., & Vaidya, S. Characterization of two distinct allyl pyrophosphatase activities from rat liver microsomes. Arch. Biochem. Biophys. 315: 393-399.

45. Qureshi, A.A., Qureshi, N., Wright, J.J.K., Shen, S., Kramer, G., Gabor, A., Chong, Y.H., DeWitt, G., Ong, A.S.H., Peterson, D. & Bradlow, B.A. (1991b) Lowering of serum cholesterol in hypercholesterolemic humans by tocotrienols (Palmvitee). Am. J. Clin. Nutr. 53: 1021S-1026S.
46. Tan, D.T.S., Khor, H.T., Low, W.H., Ali, A. & Gapor, A. (1991) Effect of a palm-oil-vitamin E concentrate on the serum and lipoprotein lipids of humans. Am. J. Clin. Nutr. 53: 1027S-1030S.
47. Wahlqvist, M.L., Krivokuca-Bogetic, Z., Lo, C.H., Hage, B., Smith, R. & Lukito, W. (1992) Differential serum responses to tocopherols and tocotrienols during vitamin E supplementation in hypercholesterolaemic individuals without change in coronary risk factors. Nutr. Res. 12: S181-S201.
48. Yu, S.G., Anderson, P.J. & Elson, C.E. (1995) The efficacy of β-ionone in the chemoprevention of rat mammary carcinogenesis. J. Agric. Food Chem. in press
49. Gould, M.N., Haag, J. D., Kennan, W. S., Tanner, M.A. & Elson, C. E. (1991) A comparison of tocopherol and tocotrienol for the chemoprevention of chemically-induced mammary tumors. Am. J Clin. Nutr. 53:1068S-1070S.
50. Tan, B. (1992) Antitumor effects of palm carotenes and tocotrienols in HRS/J hairless female mice. Nutr. Res. 12:S163-S173.
51. Ngah, W.Z.W., Jarien, Z., San, M.M., Marzuki, A., Top, G.M., Shamaan, N.A. & Kadir, K.A. (1991) Effect of tocotrienols on hepatocarcinogenesis induced by 2-acetylaminofluorene in rats. Am. J. Clin. Nutr. 53:1076S-1081S.
52. Komiyama, K., Iizuka, K., Yamaoka, M., Watanabe, H., Tsuchiya, N. & Umezawa, I. (1989) Studies on the biological activity of tocotrienols. Chem. Pharm. Bull. 37:1369-1371.
53. Guthrie, N., Gapor, A., Chambers, A.F. & Carroll, K.K. (1994) Inhibition of proliferation of MDA-MB-435 human breast cancer cells by individual tocotrienols from palm oil. Proc. AACR 35: 629 (abs.).
54. Carroll, K.K., Guthrie, N., Gapor, A., & Chambers, A.F. (1995) *In vitro* inhibition of proliferation of MDA-MB-435 human breast cancer cells by combinations of tocotrienols and flavanoids. FASEB J. 9:A686.
55. Guthrie, N., Chambers, A.F., Gapor, A. & Carroll, K.K. (1995) *In vitro* inhibition of proliferation of receptor-positive MCF-7 human breast cancer cells by palm oil tocotrienols. FASEB J. 9:A998.
56. Goh, S.H., Hew, N.F., Norhanom, A.W. & Yadav, M. (1994) Inhibition of tumor promotion by various pal-oil tocotrienols. Int. J. Cancer 57:529-531.
57. Shoff, S.M., Grummer, M., Yatvin, M.B. & Elson, C.E. 1991) Concentration-dependent increase in murine P388 and B16 population doubling time by the acyclic monoterpene geraniol. Cancer Res. 51:37-42.
58. Crowell, P.L., Chang, R.R., Ren, Z., Elson, C.E.& Gould, M.N. (1991) Selective inhibition of isoprenylation of 21-26-kDa proteins by the anticarcinogen d-limonene and its metabolites. J. Biol. Chem.266:17679-17685.
59. Schulz, S., Buhling, F. & Ansorge, S. (1994) Prenylated proteins and lymphocyte proliferation: Inhibition by d-limonene and related monoterpenes. Eur. J. Immunol. 24: 301-307.
60. Bronfen, J.H., Stark, M.J. & Crowell, P.L. (1994) Inhibition of human pancreatic carcinoma cell proliferation by perillyl alcohol. Proc. AACR 35: 431 (abs.).
61. Kawata, S., Nagase, T., Yamasaki, E., Ishiguro, H. & Matsuzawa, Y. (1994) Modulation of the mevalonate pathway and cell growth by pravastatin and d-limonene in a human hepatoma cell line (Hep G2). Br. J. Cancer 69:1015-1020.
62. Crowell, P.L., Ren, Z., Lin, S., Vedejs, E. & Gould, M.N. (1994) Structure-activity relationships among monoterpene inhibitors of protein isoprenylation and cell proliferation. Biochem. Pharmacol. 47:1405-1415
63. Hohl, R.J. & Lewis, K. (1995) Differential effects of monoterpenes and lovastatin on RAS processing. J. Biol. Chem. 270:17508-17512.
64. Kelloff, G.J., Boone, C.W., Steele, V.E., Crowell, J.A., Lubet, R. & Sigman, C.C. (1994). Progress in cancer prevention: Perspectives on agent selection and short-term clinical intervention trials. Cancer Res. 54: 2015s-2024s.
65. Ruch, R.J. & Sigler, K. (1994) Growth inhibition of rat liver epithelial tumor cells by monoterpenes does not involve ras plasma membrane association. Carcinogensis 15:787-789.
66. DeClue, J.E., Vass, W.C., Papageorge, A.G., Lowy, D.R. & Willumsen, B.M. (1991) Inhibition of cell growth by lovastatin is independent of ras function. Cancer Res. 51:712-717.
67. Beck, L., Hosick, T.J. & Sinensky, M. (1988) Incorporation of a product of mevalonic acid metabolism into proteins of Chinese hamster ovary cell nuclei. J. Cell Biology 107:1307-1316.
68. White, E. (1993) Death-defying acts: a meeting review on apoptosis. Genes & Dev. 7:2277-2284.
69. Stewart, B.W. (1994) Mechanisms of apoptosis: Integration of genetic, biochemical, and cellular indicators. JNCI 86:1286-1296.
70. Fisher, D.E. (1994) Apoptosis in cancer therapy: Crossing the threshold. Cell 78:539-542.

71. Hotz, M.A., Gong, J., Traganos, F. & Darzynkiewicz, Z. (1994) Flow cytometric detection of apoptosis: Comparison of the assays of *in situ* DNA degradation and chromatin changes. Cytometry 15:237-244.
72. Perez-Sala, D. & Mollinedo, F. (1994) Inhibition of isoprenoid biosynthesis induces apoptosis in human promelocytic HL-60 cells. Biochem. Biophys. Res. Commun. 199:1209-1215.
73. Voziyan, P.A., Haug, J.S. & Melnykovych, G. (1995) Mechanism of farnesol cytotoxity: Further evidence for the role of PKC-dependent signal transduction in farnesol-induced apoptopic cell death. Biochem. Biophys. Res. Commun. 212:479-486.
74. Clegg, R.J., Middleton, B., Bell, G. D. & White, D.A. (1980) Inhibition of hepatic cholesterol synthesis and 3-hydroxy-3-methylglutaryl coenzyme A reductase by mono and bicyclic monoterpenes administered *in vivo*.. Biochem Pharmacol 29: 2125-2127.
75. Russin, W.A., Hoesly, J.D., Elson, C.E., Tanner, M.A. & Gould, M.N. (1989) Inhibition of rat mammary carcinogenesis by monoterpenoids. Carcinogenesis 10: 2161-2164.
76. Maltese, W.A., Defendi,R., Greene, R.A., Sheridan, K.M. & Donley, D.K. (1985). Suppression of murine neuroblastoma growth *in vivo* by mevinolin, a competitive inhibitor of 3-hydroxy-3-methylglutaryl coenzyme A reductase. J. Clin. Invest. 76:1748-1754.
77. Sumi, S., Beauchamp, R. D., Townsend, C. M. Jr., Uchida, T., Murakami, M., Rajaraman, S., Ishizuka, J. & Thompson, J. C. (1992) Inhibition of pancreatic adenocarcinoma cell growth by lovastatin. Gastroenterology 103:982-989.
78. Blanock, K., Jani, J.P., Specht, S., Stemmler, N., Singh, S.V., Gupta, V., and Katoh, A. (1994) Anti-metastatic potential of lovastatin, a cholesterol lowering drug. Proc. AACR 35: 51 (abs).
79. Kawata, S., Kakimoto, H., Ishiguro, H., Yamasaki, E., Inui, Y. & Matsuzawa, Y. (1992) Effect of pravastatin, a potent 3-hydroxy-3-methylglutaryl coenzyme A reductase inhibitor, on survival of AH130 hepatoma-bearing rats. Jpn. J. Cancer Res. 83:1120-1123.
80. Shack, S., Prasanna, P., Hudgins, W. R., Liu, L., Myers, C. E. & Samid, D. (1994) Experimental therapies for malignant gliomas: Targeting the mevalonate pathway of cholesterol synthesis. Proc. AACR 35: 409 (abs.).
81. Thibault, A., Cooper, M.R., Figg, W.D., Weinberger, M.S., Tompkins, A.C., Trepel, J., Samid, D., and Myers, C.E. (1994) A phase I trial of high-dose lovastatin shows activity against malignant astrocytes. Proc. AACR 35: 226 (abs).
82. Narisawara, T., Fukaura, Y., Terada, K., Umezawa, A., Tanida, N., Yazawa, K. & Ishikawa, C. (1994) Prevention of 1,2-dimethylhydrazine-induced colon tumorigenesis by HMG-CoA reductase inhibitors, pravastatin and simvastatin, in ICR mice. Carcinogenesis 15:2045-2048.
83. Anonymous. Lovastatin Study Groups I-IV. (1993) Lovastatin 5-year safety and efficacy study. Arch. Int. Med. 153, 1079-1087.
84. Melman, M. F., Gabriel, E. D., Eskander, W. E. D. & Rao, K. N. (1987) Cholestyramine promotes 7,12-dimethylbenzanthracene induced mammary cancer in Wistar rats. Br. J. Cancer 56: 45-48.
85. Arakawa, S., Ito, M. & Tejima, S. (1988) Promoter function of carrageenan on development of colonic tumors induced by 1,2-dimethylhydrazine in rats. J. Nutr. Sci. Vitaminol. 34: 577-585.
86. Coni, P., Pang, J., Pichiri-Coni, G., Hsu, S., Rao, P. M., Rajalakshimi, S. & Sarma, D. S. R. (1992) Hypomethylation of β-hydroxy-β-methylglutaryl coenzyme A reductase gene and its expression during hepatocarcinogenesis in the rat. Carcinogenesis 13:497-499.
87. Rossiello, M. R., Rao, P. M., Rajalakshmi, S. & Sarma, D. S. R. (1994) Similar patterns of hypomethylation in the b-hydroxy-b-methylglutaryl coenzyme A reductase gene in hepatic nodules induced by different carcinogens. Molec. Carcinogenesis 10:237-245.
88. Vasudevan, S., Laconi, E., Khandelwal, M., Ackerman, P., Jones, W., Rao, P. M., Rajalakshmi, S., Marcon, N. & Sarma, D. S. R. (1994) Hypomethylation of β-hydroxy-β-methylglutaryl coenzyme A (HMG CoA) reductase gene in polyps and cancers of human colon. FASEB J. 8:A647 (abs.)
89. Wilson, D.M., Gueldner, R.C., McKinney, J.K., Lievsay, R.H., Evans, B.D.& Hill, R.A. (1981) Effect of β–ionone on Aspergillus flavus and Aspergillus parasiticus growth, sporulation, morphology and aflatoxin production. JAOCS 58:959A-961A.
90. Winterhalter, P., Herderich, M.& Schrier, P. (1990) 4-Hydroxy-7,8-dihydro-β–ionone and isomeric megastigma-6,8-dien-4-ones: new C13 norisoprenoids in quince (Cydonia oblonga, Mill) fruit. J. Agric. Food Chem. 38:796-799.
91. Charles, D.J., Simon , J.E. & Widerlechner, M.P. (1991) Characterization of essential oil of Agastache species. J. Agric. Food Chem. 39:1946-1949.
92. Krammer, G., Winterhalter, P., Schwab, M. & Schreier P. (1991) Glycosidically bound aroma compounds in the fruits of Prunus species. J. Agric. Food. Chem. 39:778-781.
93. Pabst, A., Barron, D., Etievant, P. & Schreier, P. (1991) Studies on the enzymatic hydrolysis of bound aroma constituents from raspberry fruit pulp. J. Agric. Food Chem. 39:173-175.

94. Takeoka, G.R., Flath, R.A., Mon, T.R., Teranishi, R. & Guenter, M. (1990) Volatile constituents of apricots (Prunus armeniaca). J. Agric. Food Chem. 38:471-477.
95. Buttery, R.G., Teranishi, R., Flath, R.A. & Ling, L.C. (1989) Fresh tomato volatiles: Composition and sensory studies. ACS Symp. Ser. 388:213-222. Am. Chem. Soc., Washington, DC.
96. Etoh, H., Ina, K. & Iguchi, M. (1990) 3S-(+)-3,7-dimethyl-1,5-octadiene-3,7-diol and ionone derivatives from tea. Agric. Biochem. 44:2999-3000.
97. Fujimori, T., Kasuga, R., Noguchi, M. & Kaneko, H. (1974) Isolation of R-(-)-3-hydroxy-β-ionone from Burley tobacco. Agric. Biol. Chem. 38:891-892.
98. Sefton, M.A. & Williams, P.J. (1991) Generation of oxidation artifacts during the hydrolysis of norisoprenoid glycosides by fungal enzyme preparations. J. Agric. Food Chem. 39:1994-1997.
99. Weeks, W.W. (1986) Carotenoids: a source of flavor and aromas. in Biogeneration of Aromas. ACS Symp. Ser. 317:157-166. Am. Chem. Soc., Washington, DC.
100. Voilley, A., Beghin, V., Charpentier, C. & Peyron, D. (1991) Interactions between aroma substances and macromolecules in a model wine. Lebensmittel-Wissencschaft Technol. 24:469-472.
101. Humpf, H.U., Winterhalter, P. & Schreier, P. (1991) 3,4-Dihydroxy-7,8-dihydro-β-ionone β-D-glucopyranoside: Natural precursor of 2,2,6,8-tetramethyl-7,11-dioxatricyclo[6.2.1.0]undec-4-ene (Riesling Acetal) and 1,1,6-trimethyl-1,2-dihydronaphthalene in red currant (Ribes rubrum L.) leaves. J. Agric. Food Chem. 39:1833-1836.
102. Salt, S.D., Tuzun, S. & Kuc, J. (1986) Effects of β–ionone and abscisic acid on the growth of tobacco and resistance to blue mold. Mimcry of effects of stem infection by Peronospora tabacina Adam. Physiol. Molec. Plant Path. 28:287-297.
103. Norman, S.M., Poling, S.M., Maier, V.P. & Nelson, M.D. (1985) Ionones and β-ionylideneacetic acids: Their influence on abscisic acid biosynthesis by Cercospora rosicola. Agric. Biol. Chem. 49:2887-2892.
104. Wei, C-I., Tan, H., Fernando, S. & Ko, N-J. (1986) Inhibitory effect of β-ionone on growth and aflatoxin production by Aspergillus parasiticus on peanuts. J. Food Protect. 49:515-518.
105. Feofilova, E.F. & Arbuzov, V.A. (1975) Mechanism of the action of β-ionone on carotene synthesizing enzymes of Blakeslea trispora. Microbiology 44:351-354.
106. Eslava, A.P., Alvarez, M.I. & Cerda-Olmedo, E. (1974) Regulation of carotene biosynthesis in Phycomyces by vitamin A and β-ionone. Eur. J. Biochem. 48:617-623.
107. Lewis, M.J., Ragot, N., Berlant, M.C. & Miranda, M. (1990) Selection of astaxanthin-producing mutants of Phaffia rhodozyma with β-ionone. Appl. Environ. Microbiol. 56:2944-2945.
108. Donaldson, J.M.I., Mc Govern, T. & Ladd, T.L. Jr. (1990) Floral attractants for Cetoniinae and Rutelinae (Coleoptera: Scarabaeidae). J. Econ. Entomol. 83:1298-1305.
109. Lampman, R.L. & Metcalf, R.L. (1988) The comparative response of Diabrotica species (Coleoptera: Chrysomeladae) to volatile attractants. Environ. Entomol. 17:644-648.
110. Peterson, D.M.& Qureshi, A.A. (1993) Genotype and environmental effects on tocols of barley and oats. Cereal Chem. 70:157-162.
111. Slover, H.T., Lehmann, J.,Valis, R.J. (1969) Nutrient composition of selected wheats and wheat products. III. Tocopherols. Cereal Chem. 46:635-641.
112. Piironen, V.; Syvaeoja, E.L., Varo, P., Salminen, K. & Koivistoinen, P. (1986) Tocopherols and tocotrienols in cereal products from Finland. Cereal Chem. 63:78-81.
113. Hakkarainen, R.V., Tyopponen, J.T., Hassan,.S., Bengtsson, S.G., Jonsson, S.R. & Lindberg, P.O. (1984) Biopotency of vitamin E in barley. Br. J. Nutrit. 52:335-349.
114. Barnes, P.J. (1983) Cereal tocopherols. Develop. Food Sci. 5B:1095-1100.
115. Slover, H.T. (1971) Tocopherols in foods and fats. Lipids 6:291-296.
116. Rao, M.K.G. & Perkins, E.G. (1972) Identification and estimation of tocopherols and tocotrienols in vegetable oils using gas chromatography-mass spectrometry. J. Agric. Food Chem. 20:240-245.
117. Hassapidou,M.N. & Manoukas,A.G. (1993) Tocopherol and tocotrienol compositions of raw table olive fruit. J. Sci. Food Agric. 61:277-280.
118. Jacobsberg, B., Deldime, P. & Gapor, A. (1978) Tocopherols and tocotrienols in palm oil. Consideration on their role as antioxidants and practical determinations. Oleagineux 33:239-247.
119. Goh, S.H., Choo, Y.M. & Ong, A.S.H. (1985). Minor constituents of palm oil. JAOCS 62:237-240.
120. Anderson, J.W., Gilinsky, N.H., Deakins, D.A., Smith, S.F., O'Neal, D.S., Dillon, D.W. & Oeltgen, P.R. (1991) Lipid responses of hypercholesterolemic men to oat-bran and wheat-bran intake. Am. J. of Clin. Nutr. 54:678-683.
121. Kelley, M.J. & Story, J.A. (1987) Short-term changes in hepatic HMG-CoA reductase in rats fed diets containing cholesterol or oat bran. Lipids 22:1057-1059.
122. Marlett, J.A., Hosig, K.B., Vollendorf, N.W., Shinnick, F.L., Haack, V.S. & Story, J.A. (1994) Mechanism of serum cholesterol reduction by oat bran. Hepatology 20:1450-1457.

123. Newman, R.K., Newman, C.W., Hofer, P.J. & Barnes, A.E. (1991) Growth and lipid metabolism as affected by feeding of hull-less barleys with and without supplemental beta-glucanase. Plant Foods for Human Nutrition 41:371-80.
124. Randall, J.M., Sayre, R.N., Schultz, W.G., Fong, R.Y., Mossman, A.P., Tribelhorn, R.E. & Saunders, R.M. (1985) Rice bran stabilization by extrusion cooking for extraction of edible oil. J. Food Sci. 50:361-363.
125. Saunders, R.M. (1985) Rice bran: Composition and potential food use. Food Rev. Int. 1:465-495.
126. Kahlon, T.S., Saunders, R.M., Chow, F.I., Chiu, M.M. & Betschart, A.A. (1990) Influence of rice bran, oat bran, and wheat bran on cholesterol and triglycerides in hamsters. Cereal Chem. 67:439-441.
127. Seetharamaiah, G.S. & Chandrasekhara, N. (1989) Studies of the hypocholesterolemic activity of rice bran. Atherosclerosis 78:219-223.
128. Hegsted, M., Windhauser, M.M., Morris, S.S.K. & Lester, S.B. (1993) Stabilized rice bran and oat bran lower cholesterol in humans. Nutr. Res. 13:387-398.
129. Suyama, K., Yeow, T. & Nakai, S. (1983) Vitamion A oxidation products responsible for haylike flavor production in nonfat dry milk. J. Agric. Food Chem. 31:22-36.
130. Dufour, E. & Haertle, T. (1990) Binding affinities of β-ionone and related flavor compounds to β-lactoglobulin: Effects of chemical modification. Biochim. Biophys. Acta 38:1691-1695.
131. Criqui, M.H. & Ringel, B.L. (1994) Does diet or alcohol explain the French paradox? Lancet 344:1719-1723.

7

METABOLISM OF EXOGENOUS AND ENDOGENOUS ARACHIDONIC ACID IN CANCER

Stephen D. Phinney

Division of Clinical Nutrition and Metabolism
Department of Internal Medicine
University of California at Davis
Davis, California 95616

ABSTRACT

Epidemiologic evidence in humans and controlled trials in animal models indicate that total dietary fat increases the risk of cancer. The animal evidence indicates that the greatest efficacy in promoting carcinogenesis is achieved with omega-6 fatty acids with little or no effect from either the omega-3 or monounsaturated fatty acid families. Epidemiologic studies in humans indicate a positive association between meat intake and colon cancer, but a negative association with chicken and fish. There is also a negative association between non-steroidal anti-inflammatory drug (NSAID) intake and colon cancer. Red meat is a potentially significant source of dietary arachidonic acid, which is the primary substrate for the eicosanoids whose production is blocked by NSAIDs. Thus there is a positive association between carcinogenesis and dietary intake of both the omega-6 fatty acid precursor linoleic acid and its product arachidonic acid, and a negative association with use of a drug blocking its metabolism to eicosanoids. Another potentially important factor in arachidonate metabolism is variation in its endogenous distribution. We have recently reported abnormal distribution of arachidonic acid between lipid fractions in human obesity, and parallel abnormalities in animal models of genetic obesity. This implies a potential role for variation in the endogenous distribution of arachidonic acid in the etiology of cancers which have increased incidence in human obesity. This paper addresses the role of arachidonate intake, its endogenous production, and its distribution within lipid fractions in carcinogenesis.

INTRODUCTION

Nonsmoking related cancers are on the rise in the United States population, and while there is improved survival for some forms of leukemia and lymphoma, mortality rates overall are also increasing[1]. At the same time, coronary mortality incidence is decreasing, and one

Dietary Fats, Lipids, Hormones, and Tumorigenesis, edited by Heber and Kritchevsky
Plenum Press, New York, 1996

might expect that this would contribute to a relative increase in cancer mortality. However, controlling for this factor, especially in younger age cohorts, there remains a significant increase in cancer mortality, raising concerns of an as yet undefined carcinogenic effect at work with wide exposure in our population[1].

One possible factor in carcinogenesis is total dietary fat. Up until the last decade, there has been a steady increase in the proportion of total calories consumed as fat in the United States, followed by an apparent subsequent decline over the past decade. This recent decline in dietary fat intake is based on reported food intake, which is a best suspect for judging quantitative intake.[2] Meanwhile, the disappearance of food grade fat based upon food industry statistics continues to increase. Thus it is not clear that there is an actual decline in dietary fat, as opposed to an under-reporting of this nutrient by a better educated and somewhat guilty population.

Another demographic change in fat intake has been the steady decline in both saturated fat and cholesterol intakes with a concurrent rise in the intake of vegetable source polyunsaturated fatty acids. This shift towards polyunsaturated fat intake has been the result primarily of increased high omega-6 vegetable source fats in exchange for reduced animal fats in the diet to reduce cholesterol intake. While reducing excess dietary cholesterol is probably beneficial to overall health, the increase in dietary polyunsaturates is based on the assumption that their contribution to cholesterol lowering has a net positive effect on overall health. However this purported health benefit from increasing omega-6 polyunsaturated fats above the minimum to avoid essential fatty acid deficiency has not been demonstrated in humans.

Thus it is appropriate to review the potential roles for total dietary fat, and its various biochemical fractions, in the cause of cancer; and also to consider the metabolic variables contributing to this process, with particular attention directed to the intake and distribution of arachidonic acid, an omega-6 polyunsaturate with important roles in cellular regulation.

FATTY ACID NOMENCLATURE

The primary constituents of most biological lipids are the fatty acids. Fatty acids are classed by chain length and degree of unsaturation into short chain (less than 8 carbons), medium chain (8-10 carbons), and long chain (12 carbons or longer). Unsaturation refers to double bonds between carbons, which are usually in the cis-configuration. When they occur in a fatty acid, multiple double bonds are usually spaced at every third carbon intervals. Fatty acids containing one double bond are referred to as monounsaturates, and those containing two or more double bonds are polyunsaturates. An additional classification would be those containing four or more double bonds, called highly unsaturates.

There are multiple competing methods for naming individual fatty acids. The absolutely correct system is the structurally precise IUPAC system (see Table 1). While this system precisely defines fatty acid structure, it is extremely cumbersome and provides no information about important metabolic relationships between different fatty acids. The commonly used "trivial names" provide neither structure nor metabolic information. The best method is a shorthand nomenclature consisting of the number of carbons, followed by a colon, followed by the number of double bonds, followed by the "omega position" of the first double bond. The omega position is defined as the number of carbons from the methyl (chemically inactive) end of the fatty acid, and carries with it the assumption that all subsequent double bonds will occur at 3-carbon intervals, and that all are in the cis-configuration.

As shown in Figure 1, saturates, monounsaturates, and polyunsaturates exist in families defined by the position of the first double bond (if any) from the methyl end, with

Table 1. Fatty Acid Nomenclature

Trivial	IUPAC*	Numerical+
lauric	dodecanoic	12:0
myristic		14:0
palmitic	hexadecanoic	16:0
palmitoleic	9-hexadecenoic	16:1ω7
stearic	octadecanoic	18:0
oleic	9-octadecaenoic	18:1ω9
vaccenic	9-octadecaenoic	18:1ω7
linoleic	9,12-octadecadienoic	18:2ω6
α-linolenic	9,12,15-octadecatrienoic	18:3ω3
γ-linolenic	6,9,12-octadecatrienoic	18:3ω6
Mead acid	5,8,11-eicosatrienoic	20:3ω9
DGLA	8,11,14-eicosatrienoic	20:3ω6
arachidonic	5,8,11,14-eicosatetraenoic	20:4ω6
EPA	5,8,11,14,17-eicosapentaenoic	20:5ω3
DHA	4,7,10,13,16,19-docosahexaenoic	22:6ω3

*International Union of Pure and Applied Chemists
+Shorthand nomenclature that defines fatty acid structure by the number of carbons, the number of double bonds, and the positionof the first double bond relative to the omega (methyl) end of the carbon chain. In both naming systems, double bonds are assumed to be in the *cis* configuration unless specifically stated as *trans*.

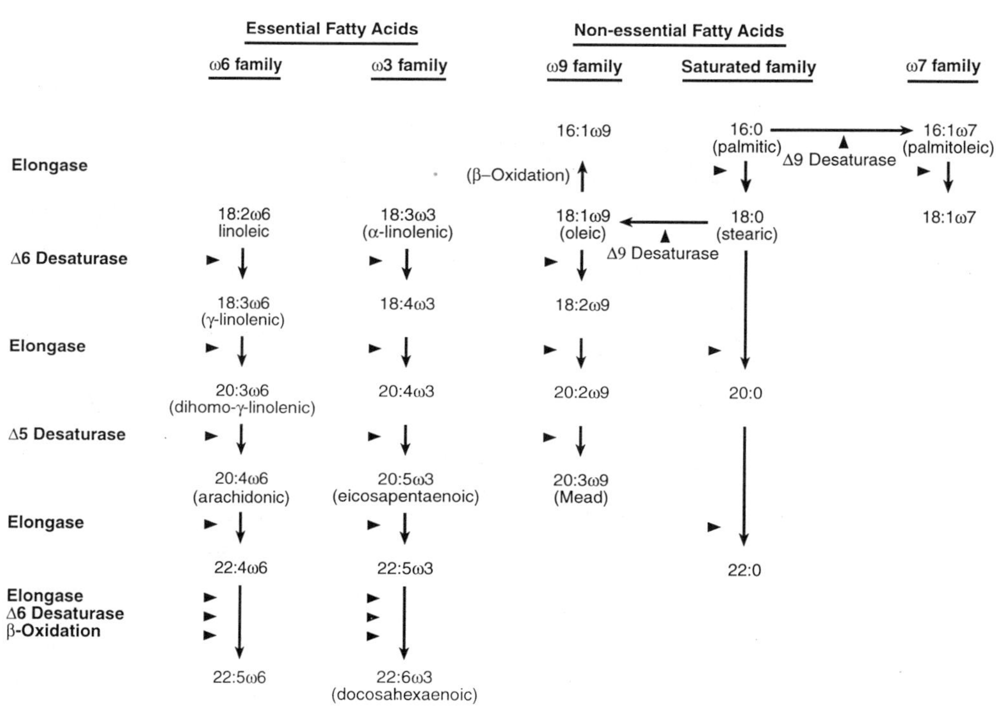

Figure 1. Nomenclature and metabolic pathways of long chain fatty acids.

metabolic interrelationships resulting from the sequential anabolic action of desaturase and elongase enzymes. The sequence of action of these anabolic enzymes is tightly constrained, and the initial delta-6 desaturase step is rate limiting. In this context, the value of the omega naming system in defining these metabolic interrelationships is obvious.

Fatty acids are relatively uncommon in their free (nonesterified) form. Most fatty acids in the human are found in triacylglycerols (commonly called triglycerides) as lipid droplets in adipose cells and to a lesser degree in other tissue cells and in circulating lipoproteins. A second important class of lipids containing fatty acids is the phospholipids, a heterogeneous mixture of predominantly diacylglycerols with a wide variety of polar headgroups attached to the remaining SN1 carbon of glycerol. The phospholipids are the predominant material in cellular and organelle membranes and at the aqueous interface of lipoproteins. Fatty acids are also commonly esterified to cholesterol, with approximated two-thirds of circulating cholesterol being in the cholesteryl ester form. A relatively small fraction of circulating and intracellular fatty acids are nonesterified, and these are bound to protein either as albumin in extracellular fluid or to fatty acid binding proteins in intracellular fluid.

FATTY ACID DISTRIBUTION

The primary focus of studies on the physiological effects of specific fatty acids has been on membranes, where the polyunsaturates tend to be bio-concentrated in the membrane phospholipids. Specific fatty acids impart specific physical properties to membranes, and participate in signal transduction through their release and subsequent metabolism[3,4]. The arachidonic acid content in membranes has been correlated with important physiological functions such as regulation of gene expression,[5,6,7] insulin sensitivity[8,9], insulin release from the pancreatic beta-cell,[10] and growth hormone action.[11] It is not yet clear, however, to what extent these effects are due to the presence of arachidonic acid in the membrane per se, as opposed to its release via phospholipases to provide substrate for enzymes yielding metabolically active compounds such as prostaglandins and leukotrienes.

One fate of arachidonate released from phospholipid is esterification to cholesterol, yielding cholesteryl arachidonate (a cholesteryl ester). This reaction is catalyzed in the serum by lecithin cholesterol acyl transferase, and it is also carried out within cells by acyl cholesterol acyl transferase. Systemically, the fate of cholesteryl ester is generally considered to be uptake by the liver and hydrolysis, although cholesterol hydrolases are also found in normal tissue and in cholesterol containing plaques in vessel walls.[12]

The arachidonic acid content of triglycerides is proportionally much lower than that of phospholipids and cholesteryl esters. Prior studies of the fatty acid composition of human adipose tissue reported arachidonic acid content as trace[13], but more recent studies using capillary gas chromatographic analysis of adipose tissue from North American adults yields a range for arachidonic acid of 0.25-0.50 weight percent. Thus a normal weight adult with 15 kilograms of total body adipose triglyceride would have an "arachidonate reserve" of between 40 and 80 grams. With weight loss, arachidonic acid is mobilized from the adipose tissue in equal proportions to other fatty acids[14,15], entering the metabolic pool at a rate of between 500 and 1,000 mg per day during total fasting or modified fasting. Under these conditions, we have reported a sharp rise in the proportion of arachidonate in serum phospholipids and cholesteryl esters, which persisted as long as the patient was actively losing weight.[16] This implies that the rate of release of adipose arachidonate during fasting is greater than the normal rate of its flux through circulating lipid pools.

SOURCES OF ARACHIDONIC ACID

As noted above, arachidonic acid can be produced endogenously from linoleic acid via the fatty acid anabolic pathway (Figure 1). However, little is known about the normal flux of substrate through this pathway in humans (i.e. the amount of linoleic acid converted to arachidonic acid each day) nor the extent to which the anabolic pathway can be suppressed by dietary arachidonic acid intake. A non-quantitative answer to this question is available from a study comparing the serum lipid fraction fatty acid compositions of vegetarian and omnivorous humans.[17] The vegan vegetarians in our study had 14% lower levels of arachidonic acid in the serum phospholipids and cholesteryl esters, suggesting that dietary intake of arachidonic acid (normally found only in animal products, and thus absent from a vegetarian diet) measurably enriches these circulating lipid fractions. A secondary observation from this study was a relative elevation in the proportion of Mead acid (20:3ω9) in the phospholipids and non-esterified fatty acids of the vegetarians. Mead acid is the non-essential polyunsaturate product that results when oleic acid transits the fatty acid anabolic pathway (Figure 1). Mead acid is pathologically elevated in overt essential fatty acid deficiency,[18] in which case its excess production is due to the lack of linoleic acid to sustain adequate arachidonate production. In our study of vegetarians, however, Mead acid was elevated compared to the omnivorous control subjects, but did not exceed the normal range. This observation implies that complete dependence upon endogenous production for one's arachidonic acid via the anabolic pathway results in a measurable increase in the production of Mead acid as a secondary product. Or, viewed from the other direction, the amount of dietary arachidonate necessary to maximally reduce the serum Mead acid level could be used as an estimate of dietary adequacy. While our knowledge of the arachidonic acid content of modern foods is incomplete, we have estimated that the daily intake of arachidonic acid by our omnivorous subjects was up to 500 mg per day.[13] Thus this value becomes a first guess for a dietary arachidonate intake that can suppress the endogenous anabolic pathway.

EXOGENOUS SOURCES OF ARACHIDONIC ACID

Arachidonic acid is found in relatively high concentration in the membrane phospholipids of terrestrial animals, and to a lesser degree in carcass fat and extracted animal fats. It is also highly concentrated in egg yolk phospholipids, especially when poultry are fed corn and soy based diets.[19] Wild fish, particularly cold water species, contain relatively little arachidonic acid, however this may change for traditional cold water species such as trout and salmon as aquaculture becomes more prevalent and terrestrial fats (seed oils and animal by-products) are introduced as feed.

ENDOGENOUS SOURCES OF ARACHIDONIC ACID

With the possible exception of premature infants,[20] arachidonic acid is not a required dietary nutrient for humans. As evidenced by the ability of humans to grow, reproduce, and live healthy lives without animal fats in their diet, adequate arachidonate can be synthesized endogenously from dietary linoleic acid precursor. Animal studies and human experience with total parenteral nutrition indicate that a minimum of approximately 1% of total daily energy intake of linoleic acid avoids overt essential fatty acid deficiency.[21] Currently, North Americans eat between 5 and 10% of their energy intake as linoleic acid, providing a copious supply of substrate to meet ongoing arachidonate needs via endogenous synthesis.

If the minimum daily requirement for linoleic acid to avoid essential fatty acid deficiency is 1% of total energy, this represents 2-3 grams per day of potential substrate for the anabolic pathway in an adult consuming 2000-3000 Kcal/day. Based on our estimates of daily arachidonate need from studies of vegetarians[17] and weight loss diets[16], we have postulated an upper limit for the daily arachidonic acid requirement of about 500 mg per day. It is not known whether this need varies with illness or with increased physical stress. Another unknown factor is the degree of interindividual variation in the human arachidonate requirement. This could vary as a result of differences in the efficiency of the uptake of arachidonic acid into phospholipids in various tissue membranes and/or its rate of release from membranes. As discussed below, however, studies of animal models of obesity provide indirect support for the concept that both membrane phospholipid arachidonate levels and the amount that is released from this pool are in part genetically determined.

ALTERED ARACHIDONIC ACID DISTRIBUTION IN OBESITY

The obese Zucker rat has a reduced proportion of arachidonic acid in liver phospholipids compared to genetically lean Zucker rats.[22,23] While others have interpreted this as a deficiency in the desaturase enzymes of the fatty acid anabolic pathway, we have reported increased arachidonic acid in the liver cholesteryl esters of the obese animals, and we also noted increased proportions of secondary products of the anabolic pathway such as 20:5ω3 and Mead acid.[24] This pattern of differences between lean and obese animals is more consistent with increased activity of the fatty acid anabolic pathway in the obese animal coupled with accelerated exchange of arachidonate from phospholipid into cholesteryl esters.[25] While the specific metabolic abnormality leading to this accelerated loss of arachidonate from phospholipids is unknown, its physiological significance in obesity has been underscored by our observation that driving arachidonic acid production via daily supplementation with gamma-linolenic acid (which bypasses the rate limiting Δ6-desaturase step) resulted in reduced food intake, reduced weight gain, and reduced percent body fat in the obese animals.[24] The same dose by the same rate of administration had no measurable effects on lean Zucker rat controls. While it is not clear that deranged arachidonic acid distribution is the primary cause of obesity in this animal model, it is clearly part of their metabolic abnormality, and it is also a characteristic of the obese genotype.

This shift in arachidonate distribution between lean and obese animals is not limited to Zucker rats. A completely different animal model of obesity, the BSB mouse, has a wide variation in percent body fat evolving from varying combinations of multiple contributing genes.[26] In this animal model, liver phospholipid arachidonate is not reduced in the obese compared to the lean. However, the obese animals do show an increased proportion of arachidonate in the cholesteryl ester pool, as well as the characteristic elevations of secondary anabolic pathway products, indicating both increased anabolic pathway activity and abnormal arachidonate distribution. Similar albeit less profound alterations in serum fatty acid distribution are also seen in human obesity.[25] In these animal models, the distortion of arachidonate distribution among lipid fractions is not as great in serum as in liver or muscle, so it remains to be determined if more prominent alterations in distribution are seen in the human liver with obesity.

IMPLICATIONS FOR ALTERATIONS IN ARACHIDONATE INTAKE AND DISTRIBUTION

Aside from providing a new perspective on genetically mediated obesity, these observations with animals and humans are the first indication that variations in omega-6 essential fatty acid intake and metabolism correlate with and/or effect the expression of a genetically mediated disease. In particular, our findings imply that there is increased release of arachidonic acid from phospholipids and uptake into cholesteryl esters associated with obesity. Since the controlled release of arachidonic acid from membranes by specific phospholipases are key components of many aspects of cell signaling and the first steps in production of eicosanoids and leukotrienes,[3,4] this would suggest that variations in the rate of arachidonate release could be an important factor in pathological processes mediated via these products.

To date, there has not been a systematic study of inter-individual variations in arachidonate metabolism and cancer risk in animals or humans. However a number of chance observations suggest this as a meaningful line of research. Willett et al[27] noted a positive association between red meat intake on colon cancer in the Nurses Health Study, but a negative association with fish intake. In an animal model, fish oil (an arachidonate antagonist), but not fish protein, reduced the incidence of pancreatic neoplastic lesions.[28] These two studies suggest that varying substrate for eicosanoid production may alter cancer risk. Another interesting observation is that NSAID use is associated with reduced colon cancer incidence,[29,30] and NSAID's block the use of free arachidonate (i.e. that which is already released from membrane phospholipid) for prostaglandin production.

If arachidonate enrichment in phospholipid and its rate of release varies between individuals due to differences in diet and genotype, it is possible that variations in arachidonate distribution are etiologic factors in human malignancy. This raises the intriguing possibility that it is the flux of arachidonic acid into tissue lipids other than phospholipids that is a primary correlate with cancer risk. Thus dietary intake of arachidonate could be a contributing factor, as well as factors affecting endogenous arachidonic acid production.

REFERENCES

1. Davis DL, Dinse GE, Hoel DG. Decreasing cardiovascular disease and increasing cancer among whites in the United States from 1973 through 1987: Good news and bad news. JAMA 1994; 271:431-437, 468.
2. Mertz W, Tsui JC, Judd JT et al. What are people really eating? The relation between energy intake derived from estimated diet records and intake determined to maintain body weight. Am J Clin Nutr 1991; 54:291-295.
3. Hartl WH, Wolfe RR. The phospholipid/arachidonic acid second messenger system: its possible role in physiology and pathophysiology of metabolism. JPEN 1990; 14:416-427.
4. Kinsella JE. Lipids, membrane receptors, and enzymes: Effects of dietary fatty acids. JPEN 1990; 14:2005-2175.
5. Clarke SD, Jump DB. Dietary polyunsaturated fatty acid regulation of gene transcription. Ann Rev Nutr 1994; 14:83-98.
6. Jurivich DA, Sistonen L, Sarge KD, Morimoto RI. Arachidonate is a potent modulator of human heat shock gene transcription. Proc Natl Acad Sci USA 1994; 91:2280-2284.
7. Thams P, Hedeskov CJ, Capito K. Exogenous arachidonic acid inactivates protein kinase C in mouse pancreatic islets. Acta Physiol Scand 1993; 149:227-235.
8. Pelikanova T, Kohout M, Valek J et al. Insulin secretion and insulin action related to the serum phospholipid fatty acid pattern in healthy men. Metabolism 1989; 38:188-192.
9. Borkman M, Storlien LH, Pan DA et al. The relation between insulin sensitivity and the fatty acid composition of skeletal muscle phospholipids. New Engl J Med 1993; 328:238-244.

10. Turk J, Gross RW, Ramanadham S. Amplification of insulin secretion by lipid messengers. Diabetes 1993; 42:367-374.
11. Nakamura MT, Phinney SD Tang AB et al. Increased hepatic Δ6 desaturase activity and differential alteration of liver and adipose fatty acid composition with growth hormone expression in the MG 101 transgenic mouse. Lipids (in press).
12. Okumura K, Kikuchi M, Matsui H et al. Abnormal arachidonate distribution in low-density lipoprotein and thoracic aorta in hyperinsulinemia. Metabolism 1995; 44:806-811.
13. Hirsch J. Composition of adipose tissue. In: Kinsell LW, ed. Adipose Tissue as an Organ. Springfield, IL: Thomas, 1962; 79-123.
14. Phinney SD, Tang AB, Johnson SB, Holman RT. Reduced adipose 18:3ω3 with weight loss by very low calorie dieting. Lipids 1990; 25:798-806.
15. Tang AB, Nishimura KY, Phinney SD. Preferential reduction in adipose tissue 18:3ω3 during very low calorie dieting despite supplementation with 18:3ω3. Lipids 1993; 28:987-993.
16. Phinney SD, Davis PG, Johnson SB, Holman RT. Obesity and weight loss alter polyunsaturated lipid metabolism in humans. Am J Clin Nutr 1990; 51:831-838.
17. Phinney SD, Odin RS, Johnson SB et al. Reduced arachidonate in serum phospholipids and cholesteryl esters associated with vegetarian diets in humans. Am J Clin Nutr 1990; 51:385-392.
18. Holman RT. The ratio of trienoic:tetraenoic acids in tissue lipids as a measure of essential fatty acid requirement. J Nutr 1960; 70:405-410.
19. Simopoulos AP, Salem N Jr. n-3 fatty acids in eggs from range-fed Greek chickens. N Engl J Med 1989; 321:1412 (letter).
20. Carlson SE, Werkman SH, Peeples JM et al. Arachidonic acid status correlates with first year growth in preterm infants. Proc Natl Acad Sci USA 1993; 990:1073-1077.
21. Wene JD, Connor WE, Den Besten L. The development of essential fatty acid deficiency in healthy men fed fat-free diets intravenously and orally. J Clin Invest 1975; 56:127-134.
22. Blond J-P, Henchiri C, Bezzard J. Delta-6 and delta-5 desaturase activities in liver from obese Zucker rats at different ages. Lipids 1989; 24:389-395.
23. Guesnet P, Boune J-M, Pascal G et al. Tissue phospholipid fatty acid composition in genetically lean and obese Zucker female rats on the same diet. Lipids 1990; 25:517-522.
24. Phinney SD, Tang AB, Thurmond DC, Nakamura MT, Stern JS. Abnormal polyunsaturated lipid metabolism in the obese Zucker rat, with partial metabolic correction by γ-linolenic acid administration. Metabolism 1993; 42:1127-1140.
25. Phinney SD. Abnormal arachidonic acid metabolism in obesity. In: Obesity:New Directions in Assessment and Management. Simopoulos A, VanItallie TB, eds. Charles Press, Philadelphia, 1995.
26. Phinney SD, Fisler JS, Tang AB, Warden C. Liver fatty acid composition correlates with body fat and gender in a multigenic mouse model of obesity. Am J Clin Nutr 1994; 60:61-67.
27. Willet WC, Stempfer MJ, Colditz GA et al. Relation of meat, fat, and fiber intake to the risk of colon cancer in a prospective study among women. N Engl J Med 1990; 323:1664-1672.
28. O'Connor TP, Roebuck BD, Peterson F, Campbell TC. Effect of dietary intake of fish oil and fish protein on the development of L-azaserine-induced pre-neoplastic lesions in the rat pancreas. J National Cancer Institute 1985; 75:959-962.
29. Peleg II, Maibach HT, Brown SH. Aspirin and nonsteroidal anti-inflammatory drug use and the risk of subsequent colorectal cancer. Arch Intern Med 1994; 154:394-399.
30. Giovannucci E, Rimm EB, Stampfer MJ et al. Aspirin use and the risk for colorectal cancer and adenoma in male health professionals. Ann intern Med 1994; 121:241-246.

8

STABLE ISOTOPES AND MASS ISOTOPOMER STUDY OF FATTY ACID AND CHOLESTEROL SYNTHESIS

A Review of the MIDA Approach*

W. N. Paul Lee

Research and Education Institute
Harbor - UCLA Medical Center
RB1, 1124 W. Carson St., Torrance, California 90502

1. INTRODUCTION

Cholesterol and fatty acids are two major components of the lipid bilayer conferring physical and functional properties to the plasma membrane. A highly regulated de novo cholesterogenesis pathway is required for membranogenesis of the normal growth and maintenance of animal cells (1, 2). Thus, rapidly growing tissues such as the brain of newborn rat pups have enhanced de novo cholesterogenesis in comparison to the adult brain (3). Conversely, slow turnover tissues such as the kidney have reduced cholesterogenesis (4). Numerous studies have shown that this regulation of lipogenesis is abnormal or absent when cells undergo malignant transformation (5, 6), and increased de novo lipogenesis has been found to strongly associate with the incidence of a number of cancers. In addition, dietary or hormonal factors which affect the regulation of lipogenesis may potentially alter the risk of development of certain cancers as suggested by the association between dietary fat intake and breast cancer risk in many epidemiological studies. The mechanism by which dietary and hormonal factors increase the risk of development of certain cancers is not known. Whether modification of dietary fat intake has any effect on the turnover of cholesterol and fatty acids or influence on the risk of developing cancer remains to be studied with the appropriate techniques.

Traditionally, radioisotopes have been used to study cholesterol and fatty acid synthesis. Incorporation of ^{14}C from 1-^{14}C - or 2-^{14}C -acetate has been extensively used to measure the relative rates of lipogenesis. The limitation of this approach is that acetate has

* Supported in part by PHS grants R01-DK46353. The GC/MS Facility is supported by PHS grants M01-RR00425 to the General Clinical Research Center, and P01-CA42710 to the UCLA Clinical Nutrition Research Unit, Stable Isotope Core.

to be converted to acetyl-CoA, the immediate precursor for lipogenesis, before incorporation can take place. The labeled acetate is potentially diluted by unlabeled acetate from metabolism and such dilution cannot be easily determined. Furthermore, the labeled acetyl-CoA may not be in equilibrium with acetyl units generated from glucose as well as beta-oxidation of long chain fatty acids. An alternative approach is the use of tritiated water in the study of lipogenesis (7-9). This approach is subject to the criticism that part of the hydrogen atoms is derived from NADPH which may not be in equilibrium with tritium in water. The large amount of tritium required virtually makes such application impossible for *in vivo* studies because of concern for radiation hazard. Recently, a novel approach employing stable isotopes and mass isotopomer distribution analysis (MIDA) has been used to study lipid synthesis and turnover. The distribution of mass isotopomers from isotope incorporation provides additional information making it possible to determine isotope incorporation and precursor enrichment in lipid synthesis in the same experiment. Thus, technical problems associated with the use of radioisotopes are potentially overcome with the use of stable isotopes and mass isotopomer analysis (10-15).

In this paper, we will review the fundamental concepts of the MIDA approach. The potential applications and limitations of MIDA will be discussed. Currently, there are three published methods of MIDA for determining fractional synthesis of lipids. These methods will be reviewed and their results compared.

2. WHAT ARE MASS ISOTOPOMERS?

A mass isotopomer is defined as a molecule of a compound with a certain number of isotope substitution. In the biosynthesis of a metabolite, the incorporation of ^{13}C or ^{2}H from labeled precursor leads to the formation of molecules with different number of isotope substitutions known as mass isotopomers. It is designated as m0, m1, m2, etc. indicating the number of isotope substitutions. Thus, m0 is the unenriched species (no isotope incorporation from its precursor), m1 is the molecules with one isotope substitution; m2, two isotope substitutions; and so on.

2.1 Units of Measure

The units of measure of isotopomer is molar fraction meaning the fraction of molecules with a certain number of substitutions. The stable isotope enrichment of a compound (ME, enrichment per molecule) which has the unit of number of isotope atoms/mole is equivalent to the specific activity of the compound in radioisotope determinations. An example of two isotopomer distributions of glucose with the same enrichment is shown in Figure 1. In sample A, there is one molecule with no isotope substitution and two molecules with one ^{13}C substitution in different locations. Therefore, m0 is 0.333, and m1 is 0.666. In sample B, m0 is 0.666 and m2, 0.333. The enrichment of each of these samples is 0.666 ^{13}C atom per molecule. If ^{14}C is used in place of ^{13}C, the corresponding specific activities of these samples will be identical to each other. The ability to quantitate the fractions of molecules with different number of isotope substitution (mass isotopomers) is one of the fundamental differences which distinguish stable isotope and radioisotope techniques.

2.2 Method of Quantitation

Stable isotope analysis of a compound consists of the determination of mass isotopomer distribution and the enrichment (ME) from its mass spectrum. Generally, it is necessary to

Sample A:

m0	o-o-o-o-o-o	33.3%
m1	o-o-o-o-o-x	33.3%
m1	x-o-o-o-o-o	33.3%

$$\text{Enrichment} = 0 \times 33.3\% + 1 \times 33.3\% + 1 \times 33.3\% = 0.666$$

Sample B:

m0	o-o-o-o-o-o	66.6%
m2	x-x-o-o-o-o	33.3%

$$\text{Enrichment} = 0 \times 66.6\% + 2 \times 33.3\% = 0.666$$

Figure 1. An example of two glucose isotopomer mixtures with the same enricchment. ^{13}C atoms are designated by (x) and ^{12}C atoms by (o).

assume that the mass spectrometer responds linearly to the concentration of the isotopomers and there is no isotope discrimination in the incorporation of tracer. The determination of mass isotopomer distribution and enrichment would be very simple if not for the presence of natural abundance of ^{13}C (and a minor contribution from oxygen-18). The natural abundance of ^{13}C (~1.1%) constitutes a large background affecting both the enrichment and mass isotopomer distribution of the compound. In Table 1, spectral data of unlabeled and two deuterium labeled palmitate specimens are shown. Spectral data are customarily presented in one of two ways, either as spectral intensities normalized to the largest peak (base peak) or intensities normalized to the total relevant current. The coefficient of variation of the individual spectral peak is subject to ion-current statistics similar to the counting statistics of radioisotope methods. The variability of peaks with low intensities will be relatively larger than those with high intensities as shown. Generally, isotopomer fraction constituting less than 0.01 (or 1%) of the compound is subject to more than 10% error. In the unlabeled palmitate, nearly 16% of the palmitate is in m1, and about 2% is in m2 due to having ^{13}C substitution for ^{12}C from natural abundance. The average number of ^{13}C substitution per molecule is given by Σmn (ME). In the case of unlabeled palmitate methyl ester, the average number of ^{13}C substitution is 0.2 (20%). Therefore, the average number of ^{13}C per atom is given by 20% divided by 19 (number of carbon atoms/molecule of methyl ester of palmitate) or 1.05%, which is approximately the natural abundance of ^{13}C. There are several methods of correction for the background natural abundance giving rise to different MIDA approaches. These different methods and their relationship to each other are reviewed in the following sections.

2.2.1 Average Mass Approach. The average mass approach to isotopic analysis by K. F. Blom (16) offers a simple method to calculate changes in isotopic enrichment (ME) of a compound from its mass spectrum. This method can be performed with very little mathematical sophistication because it has the least number of assumptions compared to other methods of background correction. For a given isotopomer distribution, the average mass (M_a) for the molecules of the compound is defined by:

$$M_a = \sum M_i(I_i/\sum I_i) \text{ (i = 0 to \%N)}, \tag{1}$$

Table 1: Normalized spectra of palmitate from HepG2 cells cultured in the presence of deuteratd water*

m/z	269	270	271	272	273	274	275	276	277	278	(M_a')obs
			(Intensities normalized against intensity of base peak, m/z 270)								
0 % D_2O	0.595	100	19.6	2.245	0.217	0.032	0.02	0.013	0.01	0.015	0.2007
c. v.	9.47%	0.00%	0.62%	0.39%	9.33%	24.12%	0.00%	43.30%	0.00%	33.30%	
4.78 % D_2O	0.673	100	74.56	35.43	11.8	2.95	0.596	0.12	0.033	0.035	0.8672
c. v.	11.37%	0.00%	2.13%	3.27%	4.01%	4.09%	6.72%	8.33%	17.32%	28.57%	
9.89 % D_2O	0.707	100	97.39	85.11	53.28	24.64	8.9	2.64	0.677	0.178	1.5844
c. v.	9.42%	1.36%	2.35%	3.38%	3.96%	4.44%	4.27%	3.79%	3.49%	22.49%	

m/z	269	270	271	272	273	274	275	276	277	278
			(Intensities normalized against sum of intensities of all peaks)							
0 % D_2O	0.0048	0.8147	0.1597	0.0183	0.0018	0.0003	0.0002	0.0001	8.15E-05	0.0001
4.78 % D_2O	0.0030	0.4421	0.3296	0.1566	0.0522	0.0130	0.0026	0.0005	0.0001	0.0002
9.89 % D_2O	0.0019	0.2677	0.2607	0.2279	0.1426	0.0660	0.0238	0.0071	0.0018	0.0005

* from reference 15

where M_i (or m/z) is the mass of the (i th) ion detected in the molecular ion group, I_i is the intensity of the ion , and $\Sigma (I_i/\Sigma I_i) = 1$ resulted from the normalization process (16). If we define isotope content of a compound (M_a') as the difference between the nominal mass (molecular weight) (M0) and the average mass (Ma), M_a' is given Eqn 2.

$$\begin{aligned} Ma' &= Ma - M0 = \\ &\sum M_i(I_i/\sum I_i) - M0\sum(I_i/\sum I_i) = \\ &\sum(Mi - M0)(I_i/\sum I_i) = \sum ni(I_i/\sum I_i), \end{aligned} \quad (2)$$

where $(I_i/\Sigma I_i)$ can be recognized as the fractional molar abundance designated as m+i of Hellerstein's notation, m_i of Lee's or Pi of Kelleher's notation (10, 13, 15). The term (Mi - M0) can be replaced by ni which is the incremental mass ranging from 0 to N which is the maximum number of ^{13}C or ^{2}H isotope substitutions possible. The isotope content (Ma′) has been calculated for the samples shown in Table 1. (Ma′)bkg is 0.2007, and the corresponding (Ma′)obs are 0.8672 and 1.5844. The calculated enrichment is therefore 0.6665 and 1.3836 respectively. The incremental average mass (^{13}C or ^{2}H atom incorporated per molecule) or enrichment (ME) is given by subtracting the Ma′ of "unenriched" compound from that observed in the labeled product. Incremental average mass is equal to

$$(\text{Ma prime})\text{obs} - (\text{Ma}')\text{bkg} = \sum ni[Ii/\sum Ii)\text{obs} - (Ii/\sum Ii)\text{bkg}] \quad (3)$$

In summary, the average mass approach provides an estimation of the enrichment or isotope content of the labeled compound. Background abundance is subtracted from the observed enrichment. However, no correction for ^{13}C natural abundance is provided for the individual isotopomers and quantitation of isotopomer distribution is not possible.

2.2.2. Excess Mass (EM) Method. The Excess Mass method grew out of the simple notion that when the number of isotope substitution is sufficiently large, the labeled isotopomer will fall outside the range of the isotopomer distribution due to natural abundance of isotopes in the natural compound known as the "isotope envelop". Under such circumstance, a simple difference between the normalized spectrum of the labeled material and that of the unenriched background would approximate the true molar fraction of the isotopomer. Another situation where this approximation would apply is the case when isotope incorporation is very low. Excess mass (EMi) is defined by Hellerstein as the difference between the fractional molar abundance of the labeled compound (Ii/ Σ Ii)obs and that of the background (Ii/ Σ Ii)bkg (10). The EM's corresponding to data of Table 1 are shown in Table 2. It can be shown that Σ ni $\times$ EMi is identical to the incremental average mass given by the equation Σ ni $\times$ ((Ii/ Σ Ii)obs - (Ii/ Σ Ii)bkg). In that sense, the EM method is a method for background subtraction in the calculation of enrichment. Thus, the EM approach is a method for determining the average number of isotope incorporated per molecule (ME). It will provide an "EM" distribution which is not equivalent to the corrected isotopomer distribution. In practice, there is substantial overlapping of the labeled and unenriched spectra. EM distribution is usually not linearly proportional to the labeled isotopomers present.

2.2.3. Multiple Linear Regression Method. It is well recognized that there are three separate components to any observed spectrum: the contribution of the derivativizing reagent, natural ^{13}C abundance and the substitution of labeled for unlabeled atoms. These components can be separated in a stepwise fashion using multiple linear regression analysis (17). In the multiple linear regression method, the contribution of the derivative component

Table 2. Mass isotopomer distribution of palmitate expressed as excess mass (EM) , molar fractions (mi) and labeled fraction (mi/Σmi). The agreement between observed and theoretical mi/Σmi ratios is evidence in support of the binomial model

	EM0	EM1	EM2	EM3	EM4	EM5	EM6	EM7	EM8
4.78 % D_2O	-0.3726	0.1699	0.1383	0.0504	0.0128	0.0025	0.0004	6.44E-05	3.25E-05
9.89 % D_2O	-0.5470	0.1011	0.2096	0.1409	0.0657	0.0237	0.0070	0.0017	0.0004
	m0	m1	m2	m3	m4	m5	m6		ME*
4.78 % D_2O	0.5413	0.2978	0.1215	0.0323	0.0061	0.0008	0.0001		0.6671
c. v.	1.55%	1.45%	2.34%	3.19%	4.66%	10.55%	41.61%		2.16%
9.89 % D_2O	0.3256	0.2555	0.222	0.1251	0.0508	0.016	0.004		1.382
c. v.	3.3	1.3	1.54	2.13	2.72	2.38	2.17		1.78%
		m1/Σmi	m2/Σmi	m3/Σmi	m4/Σmi	m5/Σmi			
4.78 % D_2O observed		0.6494	0.2649	0.0704	0.0133	0.0017			
(17, 0.0478) theoretical		0.6569	0.2638	0.0662	0.0116	0.0015			
9.89 % D_2O observed		0.3794	0.3297	0.1858	0.0754	0.0238			
(17, 0.0989) theoretical		0.3849	0.3380	0.1855	0.0713	0.0203			

* ME = Σ mn Eqn 4.

is first removed from the spectral data. Then the contribution of background natural abundance to the isotopomer distribution is algebraically subtracted using multiple linear regression to arrive at a mass isotopomer distribution due to incorporation of labeled precursor only. The detail of this method has been presented elsewhere (17). The mass isotopomer distribution (m_i) thus calculated will be proportional to the concentration of the isotopomers. It can also be shown that the average number of isotope substitution per molecule (ME) calculated by

$$ME = (\sum mn) \tag{4}$$

is equivalent to the incremental average mass of the average mass approach (16). Since the multiple regression analysis provides the "best" estimate of the true spectral distribution, the difference between ME by the regression method and the average incremental mass is the residual error of the multiple linear regression analysis. It reflects the deviation of the observed from the idealized spectrum as determined by this statistical analysis. The residual difference is a measure of the quality of the spectral data depending on the ion intensities, fragmentation pattern and linearity of response of the mass spectrometer. Any agreement between the incremental average mass and the calculated ME demonstrates that the spectral data fits that of the theoretical model of isotopomer formation from ^{13}C. In addition to providing an estimate of ^{13}C enrichment, the multiple linear regression approach also provides an estimation of corrected isotopomer distribution (m_i) not offered by the other methods. A fundamental difference between EMi and m_i arises from the fact that by definition, $\Sigma EMi = 0$ while $\Sigma\, m_i = 1$ (i =0 to N). Therefore, m_i is proportional to isotopomer concentration while EMi is not as illustrated by the difference between EM1 and m1 in the two examples of Table 2. This difference will affect the subsequent calculation of precursor enrichment and fractional synthesis in MIDA to be discussed below.

3. MASS ISOTOPOMER MODEL OF BIOPOLYMER SYNTHESIS

3.1 The Formation of Isotopomer in Biopolymer Synthesis

In the study of lipogenesis using MIDA, the distribution of isotopomers due to tracer incorporation is modeled after the synthesis of a polymer (10, 13, 18). In the absence of ^{13}C natural abundance, isotope incorporation introduced by the condensation of labeled precursor results in the formation of molecules with different number of isotope substitutions (isotopomers). The likelihood of finding a certain number of substitutions in a molecule with N possible positions for substitution is analogous to the situation of finding a certain number of success in N attempts given the probability of a success is p. Thus, the distribution of isotopomers follows that of a binomial distribution (10, 18). We have previously shown that the fraction of molecule with a certain number of substitution (m_i) is given by the coefficient of the binomial expansion:

$$m_i = (N)!/[(N-1)! \times i!] \times p^i q^{N-i}, \quad (5)$$

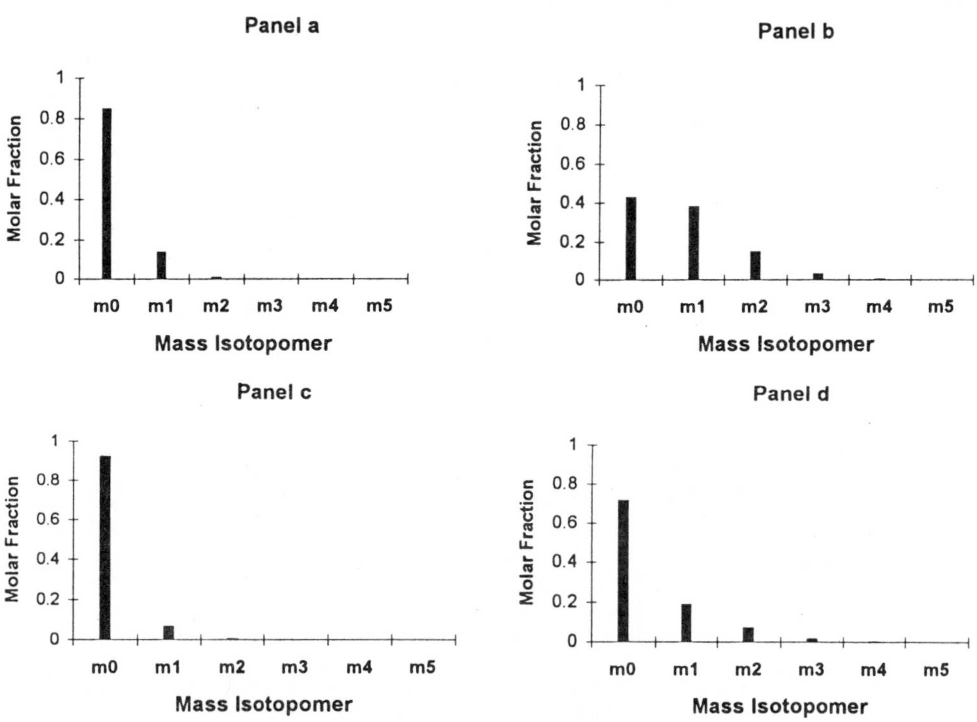

Figure 2. Distributions of mass isotopomers in palmitate due to isotope incorporation. Panels 2a and 2b are two isotopomer distributions in palmitate synthesized from two different 2-^{13}C-acetyl-CoA enrichment (2% and 10%) assuming no background ^{13}C abundance. It can be seen that the average number of substitutions is higher and the isotopomer distribution is shifted to right corresponding to a higher precursor enrichment. The effect of mixing the newly synthesized palmitate of 2a and 2b with 50% of unlabeled palmitate is shown in panels 2c and 2d. The molar fractions of labeled mass isotopomers (m1, m2, m3, etc.) relative to the unlabeled isotopomer (m0) is changed by the addition of unlabeled palmitate. However the molar fractions of labeled mass isotopomers (m1, m2, m3, etc.) relative to each other and thus the mass isotopomer ratio (e.g. m2/m1) is not affected.

where N is the maximum number of substitution possible and p the precursor enrichment. The average number of isotope substitution in the newly formed molecules will depends on the number of repeating units (N) and the precursor enrichment (p) given by N×p. The average number of isotope substitution will increase as the average mass of the compound increases corresponding to a higher precursor enrichment as illustrated by Figures 2a and 2b. Therefore, it is possible to infer the precursor enrichment from a given distribution of mass isotopomers of the product.

When the newly formed labeled compound is sampled together with some preformed (unenriched) compound, the dilution of newly synthesized molecules by the pre-existing unenriched product affects the relative distribution of the unenriched to the labeled isotopomers in this sample (Figures 2c and 2d). Equation 5 is modified to reflect this dilution and becomes:

$$m_i = \text{FSR x (N)!/ [(N} - 1)!\text{ x i!] x } p^i q^{N-i} . \tag{6}$$

It is important to note that the factor of FSR can be removed by taken mass isotopomer ratios such as m2/m1 or mi/Σmi with m0 excluded. The average isotope incorporation (MEexp) will depend on the enrichment of the precursor and the fraction of the newly synthesized molecules (fractional synthesis, FSR) given by:

$$\text{MEexp} = \text{N x p x FSR} . \tag{7}$$

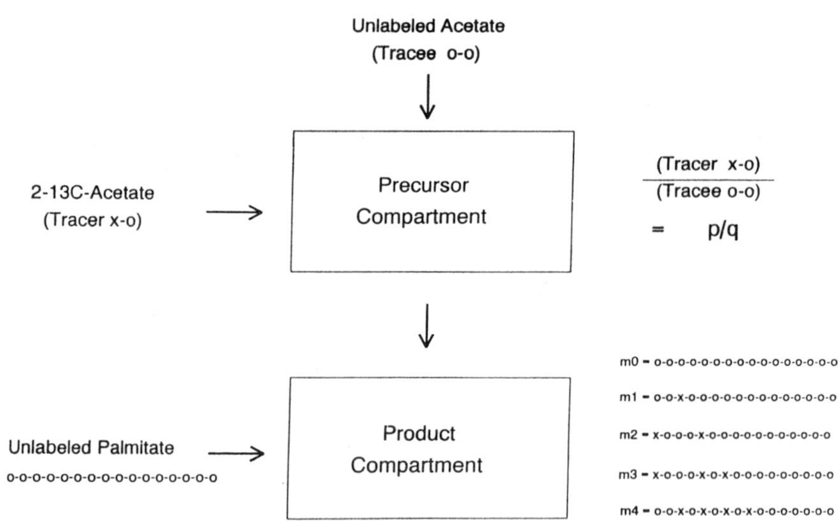

Figure 3. The biological tracer dilution model showing a single mixing compartment and a sampling compartment. Labeled precursor is first mixed with unlabeled precursor (tracee) in a single common precursor compartment. In this example of lipid synthesis, 2-^{13}C-acetate is the labeled precursor. ^{13}C atoms are designated by (x) and ^{12}C atoms by (o). The final enrichment under steady state condition is shown as (x-o)/(o-o) which is equal to p/q and p+q = 1. The newly synthesized product containing ^{13}C atoms exists as a mixture of labeled isotopomers (mi), with a distribution such that mi = N!/(i!(N-i)!) × $q^{(N-i)}p^i$. The newly formed product is then mixed with "unlabeled" molecules in the product compartment from where samples are obtained. It is assumed that the mixing in these compartments are complete and that the enrichment in the precursor compartment remains constant throughout the experiment.

3.2 The Biological Tracer Dilution Model

In the biosynthesis of lipids, isotopic label introduced by the precursor is assumed to be diluted at two points in the synthetic pathway (Figure 3). There are two necessary assumptions that there is only one precursor pool and that the labeled precursor is completely mixed with its correspondent tracee such that the enrichment of the precursor remains constant throughout the experiments (Figure 3). Tracer dilution at the level of the immediate precursor and dilution of the labeled product by pre-existing unlabeled compound at the site of sampling can have different impact on the isotopomer distribution of the product. The dilution of the labeled precursor affects the precursor enrichment and consequently the distribution of labeled isotopomers, whereas, product dilution affects the relative contribution of the unlabeled and the labeled isotopomers, and the final isotope incorporation (average number of isotope per molecule) (see Figure 2). When the necessary assumptions are met, the equations (Eqn 5 and 6) for the distribution of mass isotopomer (mi) and enrichment will apply.

3.3 Calculation of Model Parameters

There are at present three published methods for the estimation of tracer dilution in the form of precursor enrichment (p) and tracee dilution in terms of fractional synthesis rate (FSR) from spectral data depending on the method of data reduction used.

3.3.1 Excess Mass (EM) Ratio Approach. It has been shown that dilution at the tracee level (dilution by unenriched product) has no effect on the excess mass ratio. Therefore, the excess mass ratio in a product can be used to estimate precursor enrichment. In the EM method, the precursor enrichment (p) is estimated from an excess mass ratio which is usually EM2/EM1, but it can also be estimated from any other ratio such as EM2/(EM1+EM2+EM3) (10, 11, 19). Since the excess mass ratio (e. g. EM2/EM1) is a complex function of the precursor enrichment, the precursor enrichment can only be estimated graphically when the excess mass ratio is known. The next step of the method is to calculate the expected enrichment and excess mass of the newly synthesized molecules (Ap*) from the precursor enrichment (p) using an empirical equation obtained from regression of simulated data.. Fractional synthesis is then determined by dividing the observed excess mass (EM1) by Ap* (10). In this method, two theoretical curves or tables, one for the excess mass ratio and the other for Ap*, are needed for the determination of precursor enrichment and fractional synthesis.

3.3.2 Non-linear Regression Method. Another method for the estimation of precursor enrichment and fractional synthesis using non-linear regression analysis was devised by Kelleher (13). Instead of modeling the distribution of isotopomer due to isotope incorporation from labeled precursor, the observed spectrum is analyzed according to a non-linear model of tracer incorporation and the incorporation of ^{13}C from natural abundance. Natural abundance is not subtracted from the observed spectrum in this method. The step of background correction becomes part of the non-linear regression modeling process. The intensity of an observed spectral peak is a non-linear function of FSR and p. Therefore, FSR and p can be estimated from this system of non-linear equations of m_i using non-linear regression analysis.

3.3.3 Consecutive Mass Isotopomer Method. The advantage of the multiple linear regression analysis is that the contribution of ^{13}C natural abundance to the isotopomer distribution and the enrichment is properly subtracted (17). The isotopomer distribution calculated using this method represents the distribution of isotopomers due to isotope incorporation alone. Under such a condition, the mass isotopomer distribution is related to the precursor enrichment as described by Eqn 5. We have previously shown that the consecutive mass ratio mi/mi-1 is a function of the precursor enrichment ratio (p/q) where p is the enrichment of the precursor and q = 1 - p , and N (15, 18). The relationship between consecutive mass isotopomer ratio and precursor enrichment and N can be derived from equation 5. $m_i/m_{i-1} = [(N-i)/(i+1)] \times p/q$. When deuterated water is used in these studies the precursor enrichment (deuterium) enrichment can be precisely determined. The consecutive mass isotopomer ratio can be used to determine N which is the asymptotic maximum incorporation number for that particular product. Conversely, when the asymptotic maximum number of substitution (N) is known as in ^{13}C acetate studies, precursor enrichment (p) can be estimated. Since precursor enrichment is determined from the consecutive mass isotopomer ratio, we will refer to this method as the consecutive mass ratio method. Fractional synthesis (FSR) can be calculated from p by comparing the observed molar enrichment (ME) with the expected enrichment which is given by N×p. Eqn 5 and Eqn 6 suggest an alternative method which is to compared the fractional molar abundance of an isotopomer with the expected molar fraction. When m_i Eqn 6 (observed m_i) is divided by m_i of Eqn 5 (expected m_i), the quotient is FSR.

4. COMPARISON OF THREE METHODS

4.1.1 Residual Analysis. In the following section, we will use residual analysis to evaluate the three MIDA modeling methods. The residual between the enrichment predicted from the MIDA model and that calculated by the average mass method will be used to evaluate the accuracy of the particular MIDA approach. Numerically, the observed enrichment is calculated using the average mass approach. The predicted isotope incorporation is calculated from the model parameters of precursor enrichment (p) and fractional synthesis (FSR) according to Eqn 7. For example, 2-^{13}C -acetate is used to tracer palmitate synthesis, and the precursor enrichment is estimated to be 2.5% and the FSR 35%. Since there are eight acetyl unit incorporated in the synthesis, the incorporation in terms of average ^{13}C atom per molecule of the newly synthesized palmitate would be 2.5% ^{13}C atom per acetyl unit multiplied by 8 acetyl units per molecule of palmitate, or 0.20 ^{13}C / molecule of palmitate. Since 35% of the palmitate molecules are new and labeled, the average ^{13}C per molecule of the product (mixture of labeled and unlabeled) would be 0.07 ^{13}C per molecule. The expected enrichment is then compared with the enrichment calculated from the spectral data. A small residual would suggest that all necessary assumptions are met and the spectral data are properly processed. Conversely, a large discrepancy between the theoretical and the observed isotope enrichment or incorporation would suggest either that the model assumptions are not valid or that the spectral data acquisition and reduction method is not optimal. The comparison between the expected and the observed isotope incorporation for each of these three MIDA methods is presented below.

4.1.2 Residual Analysis of Selected Data. For the purpose of the discussion, we designate the three methods by their characteristic features, namely, the EM method, the non-linear regression method and the consecutive mass isotopomer ratio method. The selected set of data from three isotopomer analysis methods and the average mass calculation on the data is shown in Table 3. The normalized spectral data for m/z 270 (molecular ion of palmitate methyl ester) to m/z 279 are shown. Isotope content (Ma') was calculated for each set of spectral data using the average mass approach (Eqn 2). The model parameters corresponding to the spectral data are reproduced in Table 4, and the average incorporation's

Table 3: Selected Spectral Data of Palmitate Methyl Ester from Three Published Methods of Isotopomer Analysis

m/z -->	270	271	272	273	274	275	276	277	278	279	Ma'
					Fractional Molar Abundance						
EM Method (2-^{13}C-Ac) [a]											
Sample Label											
bkg #1	0.8222	0.1597	0.0181								0.1959
B335 8hr	0.7854	0.1833	0.0313								0.2458
bkg #2	0.8254	0.1577	0.0169								0.1915
0.06, 20%	0.7619	0.1999	0.0382								0.2763
0.08, 20%	0.7470	0.2060	0.0470								0.2999
Consecutive Mass Isotopomer Ratio Method [b]											
Sample Label											
bkg	0.8150	0.1600	0.0180	0.0020	0.0003	0.0002	0.0001				0.2048
4.78% D2O	0.4420	0.3300	0.1570	0.0520	0.0130	0.0026	0.0005	0.0001			0.8687
9.89% D2O	0.2670	0.2610	0.2280	0.1430	0.0660	0.0240	0.0071	0.0018			1.5852
Non-linear Regression Method [c]											
Sample Label											
bkg	0.8323	0.1594	0.0142	0.0002	0.0001						0.1887
1,2-^{13}C Ac	0.3825	0.0779	0.1629	0.0354	0.1488	0.0248	0.0921	0.0142	0.0425	0.0043	2.2589

a. Data from Reference 10, Table III and example of Appendix. The source of label was 2-^{13}C-acetate.
b. Data from Reference 15. Deuterated water is the source of label in palmitate synthesis in vitro.
c. Data from Reference 13, Figure 2. 1,2-^{13}C acetate (1,2-^{13}C Ac) is the source of label in palmitate synthesis.

Table 4: Intermediate Calculations Showing Background Subtraction and Model Parameters of Data Presented in Table 1

EM Method (2-^{13}C-acetate)

	EM_2/EM_1 [a]	Ap* [b]	p [c]	FSR [d]	MEexp	MEobs [e]	Δ
B335 8hr	0.5571	0.2284	0.0687	0.1035	0.0569	0.05	0.0500
0.06 20%	0.5033	0.2124	0.06	0.1989	0.0955	0.0848	0.0848
0.08, 20%	0.6221	0.2441	0.08	0.198	0.1267	0.1084	0.1084

Consecutive Mass Isotopomer Ratio Method (D_2O)

	m2/m1 [f]	N [g]	FSR	MEexp	MEobs [h]	Δ
4.78% D_2O	0.4080	17	0.821	0.6671	0.6667	0.6639
9.89% D_2O	0.8689	17	0.8218	1.3817	1.382	1.3804

Non-linear Regression Method (1,2-^{13}C acetate)

	D [i]	g(t) [i]	MEexp	MEobs [j]	Δ
1,2-^{13}C Ac	0.23	0.61	2.2448	2.070	2.070

Δ is calculated by subtracting the average mass of the "unlabeled" palmitate from the average mass of the sample. It stands for incremental average mass.
a. EM is calculated by subtracting the molar fraction of m+1 of the background spectrum from the m+1 of the sample spectrum.
b. Ap* is calculated using the regression equation y= 0.011029+4.6857x-22.156x^2 (Ref. 10).
c. p is precursor enrichment (MPE of 2-^{13}C-acetate)
d. FSR is fractional synthesis, which is calculated by dividing EM_1 by Ap*.
e. MEobs is calculated by $\Sigma\ EM_i \times n_i$.
f. m0, m1, m2 etc. stands for fractional molar enrichment of isotopomers resulted from isotope substitution during palmitate synthesis.
g. N is the number of deuterium substitutions possible during palmitate synthesis in the presence of deuterated water.
h. MEobs is calculated by $\Sigma\ m_i \times n_i$.
i. D stands for tracer dilution parameter, precursor enrichment of 1,2-^{13}C acetate, and g(t) for FSR.
j. MEobs is calculated by the average mass approach of Table 1.

(MEexp) calculated using Eqn 7 are also shown. Isotope enrichment (MEobs) was also calculated using either the EM or linear regression method. The isotope enrichment (incremental average mass or Δ) is calculated using Eqn 3. Since one does not expect any difference between the results from the average mass approach of Blom and from the EM (10, 16), the incremental average mass (Δ) and the MEobs are essentially identical to each other. However, since multiple linear regression method is a least square method, the enrichment of the calculated isotopomers (MEobs) may defer from the enrichment calculated using the average mass approach. The difference is the residual of the regression analysis. When the residual is small, the agreement between MEobs and incremental average mass would suggest that the spectral intensities are strictly linear to the concentration of isotopomers and the error from counting statistics is small.

Despite the general agreement between the predicted (MEexp) and the observed (MEobs) isotope incorporation (^{13}C or ^{2}H atom/molecule) for all three methods, there remain significant deviations of MEexp from MEobs in some of the approaches. The predicted isotope incorporation overestimated the observed by about 10% in the examples of both the EM and the non-linear regression methods. In contrast, there is virtually no difference between the expected and observed tracer incorporation in the estimation of lipogenesis using deuterated water and the consecutive mass isotopomer ratio method.

This discrepancy between MEexp and MEobs can arise from errors in spectral data acquisition or background correction; or from the deviation of the experimental system and the proposed lipogenesis model and its assumptions. To further evaluate these possibilities, the spectral data in Table 3 are transformed using the method of Lee (17) and shown in Table 5. Generally, spectral data are subject to variations due to ion current statistics, and nonlinearity of the mass spectrometer under certain circumstances. These variation will result in differences between the theoretical and observed spectrum. The transformation of the spectrum by multiple linear regression has the effect of estimating the true isotopomer distribution with the least residual. The difference between enrichment calculated from the "true" isotopomer distribution and the observed is then the residual arising from spectral variations due to counting statistics and the correction for background natural abundance. The molar enrichments shown under the column (MEcorr) of Table 5 thus represent the "best" estimates for the given spectral data. In the bottom half of Table 5, precursor enrichment and fractional synthesis were calculated using the corrected mass isotopomer distributions and the consecutive mass isotopomer method. The expected enrichment calculated from these new parameters (ME′exp) represents the result of application of the consecutive mass isotopomer method to the data used in the other two methods.

Table 6 shows the fractional residuals calculated for different processes. The total deviation of each MIDA method represented by the difference between observed (MEobs of Table 4) and expected molar enrichment (MEexp of Table 4) is shown in Column 1 of Table 6. It represents the total effect due to variation in spectral data acquisition and processing, and the lack of agreement between the experiment and the theoretical model. The difference between the corrected molar enrichment (MEcorr of Table 5) and the observed incorporation (MEobs of Table 4) reflects the variations due to spectral data acquisition and processing alone (Column 2 of Table 6). The deviation is in the range of 4-28% depending on the amount of label incorporated suggesting a major source of this variation be due to ion current statistics. When variation in the spectral data is removed by the calculation of MEcorr, the remaining difference (between MEcorr and MEexp, Column 3 of Table 6) represents error arising from the deviation from the modeling assumptions. This difference ranged from 8 to 80% of the total deviation (values of Column 3 over the corresponding values of Column 1 of Table 6). Thus, the difference between MEobs and MEexp can be attributed equally to variation in the spectral data and the lack of agreement between the experiments and the biological model.

Table 5: Reduction of Spectral Data and Calculation of Model Parameters Using Consecutive Mass Isotopomer Ratio Method

	m0	m1	m2	m3	m4	m5	m6	m7	MEcorr
				Fractional Molar Enrichment					
EM Method (2-^{13}C-acetate)									
bkg #1	0.9999	0.0008	-0.0007						0.0089
B335 8hr	0.9531	0.0376	0.0093						0.0640
bkg #2	1.0039	-0.0023	-0.0016						0.0040
0.06 20%	0.9236	0.06288	0.0135						0.0964
0.08, 20%	0.9042	0.0734	0.0224						0.1226
Non-linear Regression Method (1,2 ^{13}C acetate)									
bkg	0.9999	1.5E-05	-2.7E-05	5.54E-05	-3.5E-05				0
1,2 ^{13}C Ac	0.4672	0.0057	0.1859	0.0096	0.1701	0.0034	0.1046	0.0134	2.148

EM Method (2-^{13}C-acetate)

	m2/m1	p/q	p	FSR	ME'exp
B335 8hr	0.2485	0.0710	0.0663	0.1146	0.0608
0.06 20%	0.2152	0.0615	0.0580	0.2060	0.0955
0.08, 20%	0.3045	0.0870	0.0800	0.2056	0.1316

Non-linear Regression Method (1,2 ^{13}C acetate)

1,2 13C Ac	m4/m2 or m6/m4	p/q	p	FSR	ME'exp
m4/m2	0.9152	0.2615	0.2073	0.5698	1.890
m6/m4	0.6148	0.3074	0.2351	0.6452	2.427

p/q is determined from the equations: $m2/m1=(N-1)/2\times p/q$ or $m3/m2 = (N-2)/3\times p/q$ (Ref. 15).
FSR is calculated from $MEcorr/(N\times p)$. N is 8 for 1-^{13}C-acetate; 16 for 1,2-^{13}C-acetate; and 17 for deuterated water (Ref. 13 and 15).

Table 6. Error Analysis of Isotope Incorporation Data from Tables 4 and 5. Total error of the MIDA method is represented by (MEexp-MEobs)/(MEobs) in column 1, which can be separated into error arising from spectral data acquisition and error from modeling of the biological process as shown in columns 2 and 3. Congruence between these methods with the consecutive isotopomer ratio method is shown in column 4

	Column 1 (MEexp-MEobs)[a] MEobs (%)	Column 2 (MEcorr-MEobs)[b] MEobs (%)	Column 3 (MEexp-MEcorr)[c] MEcorr (%)	Column 4 MEexp[d] ME′exp
EM Method				
B335 8hr	13.83	28.13	11.16	0.9361
0.06 20%	12.62	13.79	1.03	0.9997
0.08, 20%	16.84	13.08	3.32	0.9625
Non-linear Regression Method				
1,2 ^{13}C Ac #1	8.44	3.7673	4.50	1.1879
1,2 ^{13}C Ac #2				0.9250

a. The difference between MEexp and MEobs represents the total error in the modeling process.
b. The difference between MEobs and MEcorr represents the error in the spectral data.
c. The difference between MEexp and MEcorr represents the error in the biological model.
d. The ratio of MEexp is a measure of the degree of congruence between the consecutive mass approach and the other two isotopomer analysis methods. The ratio will approach 1 when the methods are identical.

To show the congruence of the EM and the non-linear regression method with the consecutive mass isotopomer method, precursor enrichment and fractional synthesis are also calculated from the corrected mass isotopomer data using the consecutive mass isotopomer ratios (Table 5) (18). The congruence of these three methods can be demonstrated by comparing the ME′exp of Table 5 to the MEexp of Table 3 (Table 6, Column 4). The precursor enrichment calculated using the consecutive mass isotopomer ratios are essentially identical to those of the EM method using EM2/EM1 ratios. MEexp calculated from the published data also agreed well with ME′exp calculated using the consecutive mass isotopomer ratio method with MEexp/ME′exp ratio approaching unity (Table 6).

Since model parameters calculated using consecutive isotopomer ratio method are comparable to those derived from methods of the EM method of Hellerstein using excess molar enrichment (EM) ratio or the nonlinear regression method of Kelleher (Table 4), these MIDA methods are quite comparable to each other. However, difference among the methods remains since the MEexp/ME′exp ratio is not exactly unity. What is the origin of the difference between the experiments and the theoretical model as represented by the difference between MEexp and MEcorr ? A possible explanation can be seen in the study with 1, 2-^{13}C -acetate. In that study, the high precursor enrichment achieved gave rise to several significant isotopomer peaks in palmitate. From these peaks, two consecutive isotopomer ratios are possible. The extra information permits testing the assumption of the binomial model. If palmitate is synthesized from a constant and homogenous pool of acetyl-CoA labeled in both positions with ^{13}C, one would expect the distribution of m2, m4, m6, etc. to follow that of a binomial distribution. Consequently, the precursor enrichment calculated using m4/m2 or m6/m4 should be identical. If any of the model assumptions (of constancy of precursor enrichment or homogeneity of the precursor pool) is violated, these estimates will not agree with each other. Precursor enrichments calculated from m4/m2 and m6/m4 were different being 0.21 and 0.23. This finding strongly suggests the system deviate from the assumed binomial model as previously discussed (20). The FSR calculated using these individual values (0.57 and 0.65) subsequently do not agree with each other. However, the reported model parameters D and g(t) are bracketed by the two values of p and FSR suggesting that the non-linear regression analysis may provide an averaging of these values.

4.1.3 Conclusion. We have evaluated three isotopomer analysis methods for the study of lipogenesis by comparing the respective observed from the expected isotope incorporation. Table 6 summarizes the comparison of the predicted and observed tracer incorporation data of three published methods of isotopomer analysis for lipogenesis. In general, these methods yield similar but not identical results. The discrepancy arises from differences in spectral data reduction methods and also modeling approaches. The EM and the non-linear regression methods of Hellerstein and Kelleher do not separately analyze the contribution of error due to the spectral data analysis. Error from the spectral data acquisition and processing is lumped together with the error from modeling of the biological system. The consecutive mass isotopomer ratio approach isolates the error arising from spectral acquisition and processing from the error arising from the biological model. The residual deviation will permit the evaluation of the goodness-of-fit between the experiment and the biological model. At present, existing isotopomer data seem to suggest that the assumption for the binomial model for estimation of incorporation of isotope label from ^{13}C precursor may not be valid. If variation in precursor enrichment or compartmentalization of synthesis is a significant problem, whether linear or non-linear regression models can provide an average estimate of model parameters remains to be investigated. Until these MIDA methods are independently validated, the use of the average mass calculation and error analysis as described in this article can be used to critically evaluate the efficacy of these methods.

4.2 Sources of Error in the MIDA Approach

Deviation from the expected value was found in MIDA analysis of experimental data of studies using ^{13}C labeled precursors. Such deviations can be attributed to both error arising from the observational process such as detection of ions by the mass spectrometer, and error from the deviation of the system from the proposed model. The variation in spectral data is obviously demonstrated by the variation in the three background spectra of palmitate from three different laboratories (Table 3). The commonest source of error is from ion current statistics which inversely depends on the square root of the intensity of the peak. Other systematic errors can arise from the background noise level, tuning of the quadrupole mass spectrometers, secondary reactions after the ionization process, non-linearity of the instrument response (21), and the performance of gas chromatrographic separation. This variability can be minimized by improving the data acquisition procedure. The data reduction algorithm of Lee et al (17) can be used to provide a "best" estimate of the isotopomer distribution.

A second source of error contributing to the difference between expected and observed isotope incorporation arises from the deviation of the experimental conditions from the necessary model assumptions. In many reported studies (10-14, 22-23), the model assumptions are not verified and the validity of the model is assumed. There are two major assumptions in the isotopomer models of lipid synthesis. The first assumption is that the precursor enrichment is constant, and the second, that there is one homogeneous precursor pool. The problem of non-homogeneity of the precursor pool was recently demonstrated in the MIDA study of gluconeogenesis (24). When the necessary assumptions of Figure 3 are violated, the mass isotopomer distribution will no longer follow that of a binomial distribution resulting in an overestimation of the precursor enrichment and an underestimation of the fractional synthesis (20, 24). In the use of MIDA for lipogenesis using the non-linear regression approach, the precursor enrichment calculated from two mass isotopomer ratios yielded disparate results (Table 5). Such discrepancy is most likely the result of fluctuation

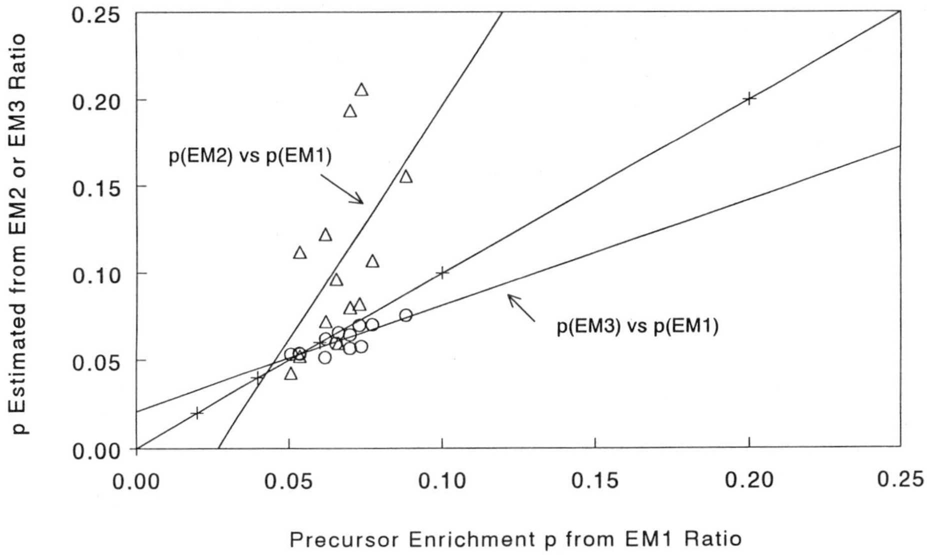

Figure 4. Correlation between precursor enrichment values calculated from human cholesterol isotopic enrichment ratios. Precursor enrichment values calculated from human cholesterol using isotopic enrichment ratio EM2/Σ EM1-4 and EM3/Σ EM1-4 are plotted against precursor enrichment values calculated from EM1/Σ EM1-4. The line of identity is included for comparison. p's are calculated from excess isotopomer ratios of subjects 1, 2, 4 and 5 from Table 2 of Reference 19. Under ideal conditions, p's calculated from any excess isotopomer ratio should fall along the line of identity. When model assumptions are violated, the p's from different calculations may not have the same correlation.

in the precursor enrichment during the incubation period. Another example of such deviations is seen in the study of cholesterol synthesis in humans reported by Neese et al (19). When the model assumptions are valid, the precursor enrichment estimated using EM1 should be identical to that calculated from EM2 or EM3 in that study. However, the estimation of precursor enrichment using EM1, EM2 and EM3 ratios gave discordant results (Figure 4). This discrepancy may be due to the presence of multiple pools of metabolic intermediates (acetoacetate, mevalonate, squalene, etc.) in cholesterol synthesis. Generally, such deviation from model assumptions will result in a difference between the observed and predicted isotope incorporation values.

Finally, when the product can arise from multiple pathways such as in stearate synthesis, disparate results can also be observed. We have found evidence for such deviations from the model assumptions in the synthesis of stearate where there are two known pathways of synthesis from chain elongation and de novo synthesis, and the newly synthesized stearate is a mixture of products from these two synthetic processes. It can be demonstrated in stearate that there is a discrepancy between N calculated from (m4/m2) and (m6/m4).

4.3 Verification of Model Assumptions

Whether the assumptions of a certain model in an experimental investigation is valid needs to be independently verified. The method of isotopomer analysis by itself does not provide that information unless the number of isotopomers measured exceeds the number of model parameters. We have previously validated the use of deuterated water in the study of lipogenesis. The validity was demonstrated by showing that the mass isotopomer distribution follows that of a binomial distribution. Consequently, any consecutive mass isotopomer ratio should give the same enrichment or N in a wide range of deuterium enrichment used. An example of this kind of verification is provided at the bottom of Table 2. The observed distribution of labeled mass isotopomer is compared to that of a theoretical binomial distribution. However, it is important to note that any agreement between the MEexp and MEobs is not necessarily a proof of the validity of the assumption. For example, in the study of palmitate synthesis using 2-^{13}C -acetate, the spectral data can only support one consecutive mass isotopomer ratio (10). The agreement between the MEexp and MEobs may simply be the result of insufficient data such that the model parameters have a unique solution. Whether any of the model assumptions is valid can only be tested when experimental data are sufficient (multiple ion peaks) to support such an analysis as was done in the case of D2O studies of lipogenesis.

4.4 Limitations

Since precursor enrichment depends on the rate of uptake and dilution of the tracer by unlabeled nutrients at the site of synthesis in cells, fluctuation of precursor enrichment may result. The fluctuation of precursor enrichment during the study would introduce a sizable error in the estimation of fractional synthesis. Whether unequal dilution or variation in precursor enrichment occurs can be checked by comparing results calculated using different mass isotopomer ratios. When a deviation from the model assumptions occurs, these ratios would yield disparate results. The non-linear regression approach of Kelleher appears to have an advantage over the others in providing an "average" estimate of the model parameters. However, the correct way of determining the "best estimate" of FSR from such data has not been developed.

5. OTHER APPLICATIONS

5.1 Condensation Synthesis vs Catenation

The theory of MIDA was originally developed for a homonucleus polymer. However, many biosynthetic processes do not follow the pattern of the condensation synthesis of a homonucleus polymer. For example, in the use of deuterated water in the study of synthesis of fatty acids, it is well known that some of the deuterium atoms are from exchange with hydrogen of water and others from reducing hydrogen of NADPH. As in the method using tritiated water, this approach is subject to the criticism that part of the hydrogen atoms is derived from NADPH which may not be in equilibrium with deuterium in water. Conceptually, the molecule can be considered as a catenation of two homonucleus polymers, one of hydrogen from water and the other of reducing hydrogen atoms (25). We have demonstrated on theoretical grounds (see below) that even in such situation and some others, the MIDA it is also applicable thus extending its usefulness in studies of lipogenesis.

5.1.1 Sterol Synthesis. In the synthesis of cholesterol as well as of fatty acids, deuterium is incorporated either through exchange with deuterium of the water molecule or through the reducing hydrogen. Since these two processes are independent of each other, we can consider the hydrogen atoms of steroid molecules to be a conjunction (catenation) of these two groups. The MIDA approach would be valid if the catenation of two binomial distribution can be modeled by a third binomial distribution. The mathematical demonstration is presented below.

Let us assume that these two binomial distributions are represented by N1, p1 and N2, p2. Their means and variances are represented by N1p1 and N2p2, and N1p1q1 and N2p2q2. Since these two distributions are independent of each other, the joined distribution would be represented by a distribution with mean being (N1p1+N2p2) and a variance of (N1p1q1+N2p2q2). It can be shown that such a distribution is equivalent to a distribution with mean of N3p3 and variance of N3p3q3 (20, 25).

The problem then becomes whether one can approximate a distribution of N3, p3 with another N4, p where p represent the enrichment of deuterium in water. The imprecision of this approximation arises from the fact that N is limited to integers only. However, when N is very large, the error arising from this approximation would be acceptably small (less than 5%). In experimental studies of lipogenesis using deuterated water, we have compared the mass isotopomer distributions of fatty acids and cholesterol to those of theoretical binomial distributions. We found excellent agreement between the observed and theoretical distributions thus experimentally demonstrating the validity of the MIDA approach in determining fractional synthesis of fatty acids and cholesterol (15, 26). Since there are other molecules which are synthesized from the condensation of two or more repeating units, such as proteins and polysaccharides, the reasoning for the use of binomial approximation in lipogenesis should readily be extended to these synthesis as well.

5.1.2 Long Chain Fatty Acids. Another instance of biosynthesis which cannot be studied with the simple binomial distribution is that of the synthesis of long chain fatty acids (LCFA) by chain elongation. The resultant LCFA product would be of two kinds depending on whether the primer is acetyl-CoA or palmitoyl-CoA. Since the LCFA product from each synthetic process would have a characteristic isotopomer distribution, the mixture can be modeled by the linear combination of these two populations using multiple linear regression analysis (27). Using this approach, we have found that LCFA from various tissues may have different proportions of its product synthesized from chain elongation and from de novo synthesis, and different fractional synthesis.

6. FUTURE DIRECTIONS

Lipid molecules exist in cells in complex forms of phosphoglycerolipids and sphingolipids. These are the molecules that confer biological properties such as surface recognition and signal transduction. The diversity of functions of these molecules are conferred by their specific head groups (phosphocholine, ethanolamine, glycerol, or inositol), the acyl groups in the second carbon position (Sn2) and the special ether or base located in the first position (Sn1). The interconversion of these lipids can be achieved by the specific phospholipases as well as enzymes responsible for synthesis of sphinganine or fatty ethers. Evidence so far has implicated sphingomyeline in the differentiation of cells, and the role of lysophospholipids as mediators in oncogene expression (28-30). Hitherto, we have discussed the application of MIDA to simple processes of lipogenesis. The study of these complex lipids would require the combined use of deuterium labeling and structural elucidation of newly synthesized lipids. This can be achieved by first isolating and quantitating the phospholipids (31), and determining their structures by mass spectrometry. Finally the turn over of each component can be determined by MIDA individually. Preliminary experiments showed that many of these complex lipids are constantly turning over by acylation and deacylation as well as exchanges of specific head groups. These specific investigations may shed new light on the role of dietary fat on the epidemiology of cancer.

7. REFERENCES

1. Srere PA, IL Chaikoff and WC Dauben. J Biol Chem 176:829-833, 1948.
2. Kandusch AA and SE Saucier. Arch Biochem Biophys 135:201-208, 1969.
3. Siperstein MD. J Lipid Res 25: 1462-1467, 1984.
4. Dietschy JM and MD Siperstein. J Lipid Res 8:97-104, 1967.
5. Habenicht AJR, JA Glomset and R Ross. J Biol Chem 255:5134-5140, 1980.
6. Huneeus VQ, MH Wiley and MD Siperstein. Proc Natl Acd Sci USA 72:50565060,1979
7. Wadke, M., H. Brunengraber, J.M. Lowenstein, J.J. Dolhun, and G.P. Arsenault. Biochemistry 12:2619-2624, 1973.
8. Jungas R.L. Biochemistry 7:3708-3717, 1968.
9. Lakshmanan M.R. and R.L. Veech. J. Biol. Chem. 252: 4667- 4673, 1977.
10. Hellerstein, M.K. J. Biol. Chem. 266:10920-10924, 1991.
11. Hellerstein M.K., M. Christiansen, S. Kaempfer, C. Kletke, K. Wu, J.S. Reid, N. S. Hellerstein, and C.H.L. Shackleton. J. Clin. Invest. 87: 1841-1852, 1991.
12. Hellerstein, M.K., C. Kletke, M. Christiansen, et al. Am. J. Physiol. 261:E479-E486, 1991.
13. Kelleher J. K. and T.M.. Masterson. Am. J. Physiol. 262 (Endocrinol. Metab. 25): E118-E125, 1992.
14. Neese, R., D. Faix, C. Kletke, K. Wu, A Ch. Wang, C.H.L. Shackleton, and M.K. Hellerstein. Am J. Physiol; 264:E134-E147, 1993.
15. Lee, W. P., S. Bassilian, Z.K. Guo, D.A. Schoeller, J. Edmond, E.A. Bergner, and L.O. Byerley. Am. J. Physiol. 266 (Endocrinol. Metab. 29) E372-E383, 1994.
16. Blom, K.F. (1988) Anal. Chem. 60:966-971.
17. Lee, W-N. P., J. Edmond, L.O. Byerley, and E.A. Bergner. Biol. Mass. Spectrom. 20:451-458, 1990.
18. Lee, W-N. P., Z.K. Guo, and E. A. Bergner. Biol. Mass Spectrom. 21: 114-122, 1992.
19. Neese, R., D. Faix, C. Kletke, K. Wu, A C. Wang, C.H. L. Shackleton, and M.K. Hellerstein. Am J. Physiol; 264:E134-E147, 1993.
20. Lee, W-N P, L.O. Byerley, S. Bassilian, H. O. Ajie, I. Clark, J. Edmond, and E. A. Bergner. Anal. Biochem. 226: 100-112, 1995.
21. Patterson, B. W. and R.R. Wolfe. Biol Mass Spectrom. 22: 481-486, 1993.
22. Neese R. A.; Schwarz, J. M.; Faix D.; Turner, S. , et al. J. Biol. Chem. 270:14452-14463, 1995.
23. Landau, B. R., Fernandez, C. A., Previs S. F., Ekberg, K., et al. Am. J. Physiol. 269: E18-E26, 1995.
24. Previs, S. F., Fernandez, C. A., Yang, D. W., Soloviev, M. V., et al. J. Biol. Chem. 270: 19806-19815, 1995.
25. Lee, W. P., Byerley, L.O., Bassilian, S., Bergner, E. A. and Ajie, H. In Modeling and Control in Biomedical System, Proceedings of the International Automatic Control (IFAC) Symposium. B. W. Patterson (ed), University of Texas Medical Branch, Galveston, TX , pp 352-353, 1994.
26. Lee, W-N. P., S. Bassilian, H. Ajie, D.A. Schoeller, J. Edmond, E.A. Bergner, and L.O. Byerley. Am. J. Physiol. 266: E699-E708, 1994.
27. Ajie, H. O., M. J. Connor, W. N. P. Lee, S. Bassilian, E. A.Bergner, and L. O. Byerley. Am. J. Physiol. 269: E247-E252, 1995.
28. Collins S. J. Blood 70: 1233-1244, 1987.
29. Dbaibo GS, Obeid LM and Hannun YA. J. Biol. Chem. 266:17762-17766, 1993.
30. Schroeder J. J., Crane H. M., Xia J., Liotta D. C. and Merrill A. H. J. Biol. Chem. 269: 3475-3481, 1994.
31. Foch J, Lees M., Stanley GHS. J. Biol. Chem. 226:497-509, 1956.

9

LIPID BIOMARKERS OF ADHERENCE TO LOW FAT DIETS

Judith M. Ashley

UCLA School of Medicine
Division of Clinical Nutition
Los Angeles, California 90024

1. INTRODUCTION

The need to lower fat intake in the U.S. has been emphasized in at least fifteen major national dietary recommendations issued since the 1970's. These include the first Senate Select Committees Dietary Goals in 1977, the Surgeon General's reports in 1979 and 1988, which laid the groundwork for the concept that changes in food intake could make a significant impact on chronic diseases, the USDA updated guidelines in 1990, and the National Academy of Sciences recommendations, as well as guidelines from the American Cancer Society and the American Institute for Cancer Research[1-6]. Although the latest periodic study of the U. S. diet, HANES III (National Health and Nutrition Examination Study, 1988-91), shows a downward trend toward an average fat intake of 34%, this still falls short of the recommended 30% or less energy from fat [7].

Studies of food intake rely primarily on dietary records, including those based on subjects' recall of recent diet (24-hour recalls), on retrospective dietary history (food frequencies) or on prospective records of food consumption (daily food records). Each of these methods is useful, but each is also fraught with potential errors of validity and reproducibility, including those of recall memory, reactive changes in usual eating habits during documentation, and under- or over-reporting[8]. Thus there is a compelling need to document adherence to low fat diets through "biomarkers" which are not dependent on self-report and to integrate these biomarkers into large-scale observational and intervention research studies.

2. THE ESTABLISHMENT OF BIOMARKERS

An ideal biomarker or combination of biomarkers of dietary fat would have adequate sensitivity to absolute fat quantity as well as to fat quality in the diet. Fat quantity refers to the absolute number of grams of fat or the percentage of energy from fat ingested in the diet. Fat quality refers to the possible change in the pattern of fats from the families of fatty acids

Dietary Fats, Lipids, Hormones, and Tumorigenesis, edited by Heber and Kritchevsky
Plenum Press, New York, 1996

including saturated, monounsaturated and polyunsaturated fats and the associated fatty acids within the families of fatty acids ingested. These biomarkers should have a time course of response which would permit their utilization in research and dietary intervention trials, including analysis in free-living and more controlled feeding conditions. They could serve as important markers to examine change over time as well as validate the association between dietary intake as measured by self-report dietary records.

To date, there are no established biomarkers of dietary fat intake. Because it has long been recognized that the amount and nature of fat ingested affects plasma cholesterol concentration, total, LDL- and HDL- cholesterol and their lipoproteins carriers have been the most common surrogates used in the past (primarily in cardio-vascular studies) as markers of high or low fat intake. However the nature of the relationship of lipoproteins to fat intake varies by individual and may reflect genetic variation more than differences in fat intake, which appear to show a real but small influence[9]. What may be more relevant as biomarkers of dietary fat intake are the fatty acids themselves, which are the basic constituents of lipids and body depots of fat. Depending on the nature and duration of the study, fatty acids in different body compartments represent important temporal relationships with dietary fat intake. In general, fatty acids in the plasma represent short term-indices (days), blood cells represent medium-term indices (months) and tissue depots represent more long-term indices (years)[10]. Thus there may be little value in measuring change in adipose tissue in a study lasting only several weeks, and free fatty acids in plasma, which are affected by what has just been eaten, may not reflect the levels of fat incorporated in cell membranes.

3. THE ROLE OF LIPIDS

Fats perform several important roles in the body and in food. They form body structures, are integral to the regulation of metabolism and provide the major source of energy. Structural lipids and depot fats are the two forms of lipids found in cells. Virtually all cell membranes, including the nerve membranes, consists partly of lipids, including cholesterol, phospholipids and sphingolipids. Adipose cells contain a large amount of triglycerides. Fats serve also as essential fatty acids (linoleic and linolenic fatty acids) which must be supplied by the diet and are the precursors for eicosanoids (prostaglandins and leukotrienes) and aid in the transport and absorption of the fat soluble vitamins (A, D, E and K). In foods, the three major fats in foods are triglycerides, cholesterol, and phospholipids.

3.1. Triglycerides

Triglycerides (triacylgylcerols) are the main form of energy in the body and are by far the principal form in which fats are eaten and stored. They are composed of three fatty acids attached by ester bonds to a glycerol. Most triglycerides are in mixed forms, containing different types of fatty acids on each glycerol molecule. Fatty acids are long chains of carbons bonded to hydrogen atoms. At one end of the molecule, called the alpha end, is an carboxyl or acid group. At the other end is a methyl group. The fatty acid chains vary in length and the degree of saturation with hydrogen. The fatty acid components of lipids are classified as short-chain (less than 6 carbons), medium-chain (6 to 10 carbons) or long-chain (12 or more carbons). While medium chain and short chain fatty acids are easier to digest and absorb, most fats from foods contain predominantly long chain fatty acids.

3.2 Fatty Acids

In saturated fatty acids, all of the bonds are single, in monounsaturated fatty acids there is only one double bond, and in polyunsaturated fatty acids there are two or more double bonds. Palmitic (16 carbons) and stearic acids (18 carbons) are the most common saturated fatty acids, while the major monounsaturated fatty acid is oleic acid (18 carbons). Plant fats are predominantly unsaturated. Olive and canola oils are good sources of monounsaturated fats, while corn, soybean, and safflower oils are good sources of polyunsaturated fatty acids. However, all fats contain a mixture of saturated and unsaturated fatty acids.

In order to make vegetable oils solid at room temperature (margarine and shortening), they are chemically processed by adding hydrogen to some of the unfilled double carbon bonds. During this saturation process, the normal position of hydrogens at the double bond known as *cis*, or same side, may be partly transposed so that the hydrogens are on the opposite sides of the double bond, resulting in *trans* fatty acids. Products containing partially hydrogenated fatty acids contain both cis and trans fatty acids.

Animals and humans have several metabolic pathways for transforming and synthesizing fatty acid. Fatty Acid Families are defined by where the carbon-carbon double bonds begin, because this placement of the first double bond makes a difference in how the body metabolizes a fatty acid. There are four Families or groups of fatty acids defined in terms of the position of the first double bond from the methyl end of the carbon chain: omega-3 (n-3), e.g. 18:3 n-3 or alpha-linolenic acid, omega-6 (n-6), e.g. 18:2 n-6 or linoleic acid, omega 7 (n-7), e.g. 16:0 n-7 or palmitoleic acid, and omega-9 (n-9), e.g. 18:1 n-9 or oleic acid. Fatty acids undergo desaturation and elongation in the body using enzymes common among the different fatty acid families, with their affinity for these enzymes following the order n-3 > n-6> n-9 > n-7. In addition, the body cannot produce carbon-carbon double bonds at the omega-3 and omega-6 positions, making linoleic and alpha-linolenic acid essential fatty acids. Arachidonic acid is a longer chain omega-6 fatty acid with 20 carbons and 4 double bonds (20:4, n-6). Fish oils contain two common longer chain omega-3 fatty acids with 20 carbons or more, eicosapentaenoic acid or EPA with 20 carbons and 5 double bonds (20:5, w-3) and docosahexaenoic acid or DHA with 22 carbons and six carbon double bonds (22:6, n-3).

The longer chain highly polyunsaturated fatty acids form a group of biologically active compounds called eicosanoids: prostaglandins, prostacyclins, thromboxanes, and leukotrienes. In essence the eicosanoids act as hormone-like compounds, regulating such vital functions as blood pressure, childbirth, blood clotting, immune responses, inflammatory responses and stomach secretions. By starting with long chain fatty acids, cells synthesize different types of eicosanoids. Arachidonic acid, which is an omega-6 fatty acid, yields one type of prostaglandin and thromboxane, while EPA and DHA, which are omega-3 fatty acids, yield another type of prostaglandin and thromboxane. For example, products made from omega-3 fatty acids tend to lower blood clotting, blood pressure, and inflammatory responses in the body, while those from omega-6 fatty acids tend to have the opposite effects.

3.3 Phospholipids

The second largest lipid component of the body is made up of phospholipids which are triglycerides in which one of the fatty acids is replaced by a phosphorus-containing substance such as phosphoric acid. Because of their strong affinity for both water-soluble and fat-soluble substances, phospholipids are effective structural materials. Large concentrations are found in combination with protein in cell membranes where they facilitate the passage of fat in and out of the cell, and in the blood where they also function in the transport of lipids. The nature of the fatty acids in phospholipids is influenced by the fatty acid content of the diet. Lecithin, or phosphatidylcholine, contains phosphoric acid and the nitrogen containing base choline. It is the most widely distributed of the phospholipids, and liver, egg yolk and soybeans are good sources. Other phospholipids such as cephalins (similar in structure to lecithin), lipontols (which contain inositol, a compound with vitamin-like activity), and sphingomyelins (which contain a complex amino alcohol in place of glycerol) are found in high concentration in nerve tissue. Even though phospholipids are found in food, the body also makes them when needed.

4. PLASMA LIPOPROTEIN CONCENTRATIONS

Changes in dietary fatty acid intake have been shown to result in changes in serum cholesterol levels. In general, the intake of saturated fatty acids has been shown to raise plasma total and LDL-cholesterol levels, whereas the intake of polyunsaturated and monounsaturated fatty acids has been shown to decrease them[11]. Of interest is the early suggestion that replacement of saturated by polyunsaturated fatty acids in the diet may lower serum cholesterol concentrations because the liver preferentially converts polyunsaturated fatty acids into ketone bodies instead of VLDL lipoprotein triglycerides[12]. However, more recent work shows that dietary fat saturation affects all aspects of lipoprotein metabolism, from synthesis to intravascular remodeling and exchanges to receptor-mediated and nonspecific catabolsim[13]. Most important may be the fact that responses to dietary fat are highly individualized and have been shown to be subject to effective feedback control mechanisms[14]. Their usefulness is also dependent on the distribution of serum cholesterol levels in the population studied[15].

Although plasma total cholesterol and LDL-cholesterol levels may respond variably depending on the type and amount of fatty acids consumed, HDL-cholesterol levels appear to be more sensitive to the total fat intake irrespective of the composition of the fatty acids[11]. A recent study which examined the effects of a 15% of energy from fat intake for one year found that the parameters most closely associated with change in dietary fat intake were HDL-cholesterol, HDL2-cholesterol and lipoprotein lipase[16]. The same author previously found a correlation with post-heparin lipoprotein lipase activity[17]. However, the need to administer intravenous heparin in order to measure lipoprotein lipase makes it usually impractical for most intervention studies. Although large amounts of n-6 polyunsaturated fatty acids has been shown to lower HDL-cholesterol levels, such a high intake of these fatty acids is not common in most studies[11]. Upon adaptation of a low fat diet, blood levels of the major apolipoproteins (A-1 and B) also seem to change in parallel with the changes in their corresponding major lipoproteins. Recent work has demonstrated that apo A-IV levels are sensitive to acute changes in dietary fat content, and can be used to monitor short term changes[18]. However, apoA-IV stabilizes in relation to a new dietary fat intake and loses its value in most long-term studies.

5. FATTY ACIDS IN FLUIDS AND TISSUES

There have been numerous published studies comparing the dietary intake of fatty acids from food with circulating lipids and tissue fatty acids, including plasma total cholesterol esters, phospholipids, and triglycerides, red blood cells, buccal (cheek) cells, and adipose depot tissue. These studies have looked at associations, both cross-sectionally and in intervention studies. The major conclusions of these studies follow our knowledge of the complex nature of fatty acid metabolism. Although fatty acids are the basic constituents of all cell membranes and can be directly measured, their tissue concentrations are modified by metabolism of the various classes and they often do not directly reflect dietary intake. In addition, the concentration of fatty acids in tissues reflects dietary intake for fatty acids that are not synthesized or formed in the body. For example, the common monounsaturated and fatty acids which the body most easily synthesizes also accounts for the majority of the total dietary fatty acid intake in the U. S.[7].

5.1 Depot Fat

It is well established that the fatty acid composition of diets strongly influences the type of triglyceride fatty acid stored in adipose cells. Fat absorbed in the intestine reaches the adipose tissue directly in the form of chylomicrons or from the liver as lipoproteins[19]. Triglycerides can also be synthesized directly in the adipose tissue from glucose.

Fatty acid profiles in adipose tissue have long been shown to generally reflect long-term dietary intake[20,21]. It has been known since the work of Hirsch in 1960 that although a small proportion of fat is rapidly available for metabolic needs, the turnover of fatty acids in adipose tissue storage is very slow[22]. When the body weight remains stable, the half-life of fatty acids in adipose tissue storage is of the order of 600 days[23]. It has been estimated that a dietary change in the proportion of different types of fatty acids would take one to two years to become apparent from the analyses of adipose fatty acids and about three years before being completely reflected as a biomarker[24]. This makes the correlation of fat biopsy of adipose tissue reflective of long periods of intake for dietary assessment[25], but perhaps too slow to be of practical use as a biomarker in short-term studies[26]. In addition, whenever a subject is losing weight, as might be the case for those following a low fat diet (see below), changes in dietary fatty acids will take much longer to appear in the tissues as fat catabolism for energy is increased[27].

Dietary polyunsaturated fatty acid intake, primarily linoleic acid, has been shown to correlate well with tissue concentrations of linoleic acid or the order of r=0.50-0.83 in subjects of differing genders, ages, and methods of dietary assessment[17,28-31]. One investigator also found a correlation of r=0.62 between the linoleic acid to saturated fatty acid ratio of fat tissue and diet[25]. Large differences were also commonly reported in tissue linoleic acid concentrations upon dietary modification with polyunsaturated fat[17]. Because linoleic acid is the most common polyunsaturated fatty acid found in the diet, it would be expected to fall as the fat level of the diet is reduced. It is also not unexpected that correlation coefficients found with monounsaturated fatty acids and saturated fatty acids are generally low, most being in the range of r=0.15-0.30[10,17.] Some changes in oleic acid concentrations have been reported with interventions using large amounts of oils containing this fatty acid, but this is not consistently seen[28]. The changes seen with oleic acid is also smaller than those seen with polyunsaturated fatty acids. Long-term ingestion of large amounts of n-3 fatty acids has been reflected by the fatty acid composition of aspirated adipose tissue samples[32]. However, n-3 adipose tissue fatty acids also need to be further studied in relation to lower intakes

Although it was once considered that subcutaneous fat taken from different parts of the body would not show substantial differences within the same subject, it is now known that the relative percentages of fatty acids varies from tissue to tissue, partly due to the competition between fatty acids, selective uptake and potential differences in in-situ fatty acid modification[33]. Thus it is important to have baseline assessments in each tissue against which to compare change. In determining markers of dietary fat intake, the feasibility and subject acceptability of the subcutaneous fat biopsy needs consideration. Some investigators find that subjects are more apprehensive about a fat biopsy than a venipuncture[32,36] ,while others report no major complaints [34,35].

5.2 Phospholipid Fatty Acids in Cell Membranes

As essential constituents of cell membranes, phospholipid content varies by the type of cell, its function and the type of membrane[37]. These structural lipids appear more resistant to compositional changes than stored triglycerides[38]. The fatty acid composition of erythrocyte membranes was suggested as a biomarker for investigating dietary fat intake over 30 years ago by Horowitt[39] and by and by Farquhar and Ahrens[40]. The method has many practical advantages over adipose tissue sampling, since it is usually more acceptable to subjects and does not require specialized training or equipment. The main components in red blood cell membranes are phospholilpids, which represent 97% or more of the total, with glyerides, cholesterol esters and free fatty acids accounting for the remaining 3%.

The changes in red cell membrane fluidity can also be a measure of dietary fat intake[41-43]. In a recent study, the anisotropy fluorescence of the probe 1,6-diphenyl - 1,3,5 - hexatriene (DPH) was measured in the low density lipoprotein (LDL) fraction of plasma[44] . Increased fluidity was noted when subjects reduced their fat intake from 40 to 20 percent of energy at either a polyunsaturated to saturated (P/S) ratio of 1.0 or 0.3. Platelets have also been used to study fatty acid composition[45]. However, turnover of fatty acids is not the same in platelets as in red blood cells, and the fatty acid compositions are different. Analyses of platelet membranes has been used more commonly to investigate the specific effect of fatty acid composition as a possible modifier of platelet function in relation the formation of thrombosis[46,47].

A novel method has been used by McMurchie to look at the fatty acid composition of tissues from buccal (cheek) cell membranes obtained non-invasively by collecting several mouth washing from subjects[48,49]. Because the turnover of human cheek cells is approximately five days, the type of fatty acids incorporated are an indicator of dietary intake over a short period of time. Results from McMurchie suggest that change from a low to a high polyunsaturated to saturated fat ratio in the diet resulted in a in a 36% increase in the proportion of linoleic acid in the cheek cell phospholipids within a six week period. A subsequent decrease did not lead to a similar reduction, suggesting that some mechanism regulates the incorporation of linoleic acid close to an optimum level or that a longer period or a much lower intake is required to observe any change.

5.3 Fatty Acids in Plasma

Fatty acids are present in plasma mainly as components of triglycerides, cholesterol esters and phospholipids. The fatty acid composition of triglycerides is closely related to dietary intake on a short-term basis and has been used to monitor dietary changes in subjects following special diets[50]. A diet rich in polyunsaturated fatty acids causes an increase in linoleic acid in triglycerides within a week. As a subject continues to follow the diet, the linoleic acid levels reach a plateau and remain fairly constant. Some investigators find that a better indicator of medium-term intake is the fatty acids linked to cholesterol esters and phospholipids. In an analysis of fatty acid compositions of serum phospholipids in fishing communities versus inland inhabitants, the percentage of omega-3 polyunsaturated fatty acids have been found to be significantly higher in the fishing villages, accurately reflecting differences in fat intake[51]. However it has also been shown that the percent linoleic acid in total plasma lipid was as good an indicator of dietary intake as the proportion in phospholipid and cholesterol esters[52]. This may be an important consideration since the laboratory measurement of serum fatty acids is far less time consuming and can accommodate a larger number of samples[53].

While the concept that fatty acid profiles in plasma might be used to follow dietary fat intake is not new, the importance of performing quantitative analysis providing absolute concentration data has recently been established. While a qualitative analyses or relative percentage method can be used for the much less complex problem of marking the intake of significant amounts of distinct single classes of exogenous dietary fats such as fish oils, it does not distinguish individuals consuming a low fat diet from those consuming a high fat diet. Moreover, to the extent that the quantities of both saturated and unsaturated dietary fatty acids are decreased in the same proportions, the relative percent composition of plasma fatty acids is less reflective of the change in dietary fat intake than quantitative total fatty acid analysis. One recent study examined in detail the effect of various diets on serum fatty acids which were consumed for six months. In comparison with baseline values, the low total fat diet of 28% of energy from fat showed an increase in serum cholesterol ester concentration of C16:1, C18:1 and C20:5 and a decrease in linoleic acid[54]

Blackburn and co-workers conducted studies in 12 free-living subjects to determine quantitative and qualitative plasma phospholipid (PL), free fatty acid (FFA), triglyceride (TG) and cholesterol ester (CE) fatty acid variations over time (0.8, and 22 months) and to correlate these fatty acids with dietary intake[55]. Self-reported dietary intake did not vary over time and quantitative phospholipid fatty acids (umol/L) also did not change significantly over the same period. Most fatty acid variations were quantitative occurring in the FFA and CE fractions. This study was important in pointing out the importance of using quantitative analyses, rather than depending on qualitative analyses of fatty acids.

Based on these results demonstrating a relationship between plasma quantitative fatty acids and dietary fat, a larger study was undertaken as part of the WINS (Women's Intervention Nutrition Study) for prevention of breast cancer recurrence in women with early resected breast cancer[56]. The intervention was a field study encouraging the intake of a diet in which 15% of total calories was derived from fat. Women ages 50 to 75 with stage II breast cancer who consumed ≥ 30% of calories as fat at baseline were recruited. The low fat intervention consisted of biweekly individual sessions for three months followed by monthly sessions thereafter. Individual food intakes were calculated based on 4 day food records collected at baseline and after three months of dietary intervention. Changes in plasma total fatty acids were measured by gas liquid chromatography following extraction of plasma samples with chloroform:methanol and transmethylation by a modification of the method of LePage and Roy[53]. Changes in quantitative fatty acid analyses performed on 43 subjects randomized to the intervention group at baseline, 6 months, and 12 months are shown in table 1 below.

It should be stressed that this measurement of quantitative fatty acids is quite different than the traditional measurement of relative percent of fatty acids. The relative percents are expected to change markedly when the P:M:S ratio changes but not when the quantity of fat

Table 1. Changes in quantitative plasma fatty acids in low-fat group quantitative fatty acids (umol/L)

Fatty acid	Baseline	6 Months	12 Months
14:0	108 ± 11	96 ± 11	142 ± 41
16:0	1437 ± 118	1187 ± 69*	1290 ± 169
16:1 n-7	149 ± 16	145 ± 12	198 ± 33
18:0	499 ± 52	358 ± 17*	374 ± 38*
18:1 n-9	1536 ± 182	1005 ± 79*	1175 ± 156*
18:2 n-6	2046 ± 176	1442 ± 110*	1519 ± 179*
18:3 n-3	62 ± 6.3	39 ± 7.7*	42 ± 9.3*
20:3 n-6	104 ± 9.6	87 ± 7.7	93 ± 23
20:4 n-6	631 ± 54	522 ± 31*	484 ± 48*
SUM	6567 ± 554	4874 ± 292*	5364 ± 509*
SFA	1920 ± 236	1547 ± 82	1749 ± 241
MUFA	1702 ± 193	1159 ± 86*	1325 ± 211
PUFA	2942 ± 236	2168 ± 145*	2291 ± 213*

All values mean ± SEM, n=43
* values with asterisk significantly different from baseline
$p < 0.05$ by ANOVA.

Table 2. Changes in relative % plasma fatty acids in low-fat group relative percent fatty acids (%)

Fatty acid	Baseline	6 Months	12 Months
14:0	1.3 ± 0.18	1.7± 0.21	2.1 ± 0.38*
16:0	20.6 ± 0.64	22.7 ± 0.59*	23.0 ± 1.04*
16:1 n-7	2.11±0.15	2.77 ± 0.21	2.95 ± 0.55
18:0	7.83 ± 0.27	7.70 ± 0.18	7.46 ± 0.33
18:1 n-9	22.9 ± 0.99	20.7 ± 0.64*	21.8 ± 0.82
18:2 n-6	31.4 ± 0.74	29.7 ± 0.96	29.4 ± 1.60
18:3 n-3	1.04 ± 0.13	0.74 ± 0.15*	0.80 ± 0.24
20:3 n-6	1.74 ± 0.13	1.97 ± 0.14	1.93 ± 0.24
20:4 n-6	11.1 ± 0.61	12.0 ± 0.51	10.6 ± 1.00
SUM	100%	100%	100%
SFA	29.7±0.60	32.2 ± 0.83*	32.6 ± 1.38*
MUFA	25.0 ± 0.97	23.5 ± 0.63	24.7 ± 1.18
PUFA	45.3 ± 0.91	44.4 ± 0.99	42.7 ± 2.00

All values mean ± SEM, n=43
* values with asterisk significantly different from baseline
$p < 0.05$ by ANOVA.

is changed with little alteration in the P:M:S ratio. These data are shown below in table 2 for the same study to illustrate the differences found by quantitative fatty acid analysis as compared to a simple analysis of relative percent of fatty acids on the chromatogram.

As demonstrated, the use of internal and external quantitative standards (13:0 and 15:0) allows the quantitation of individual fatty acids in umol/L. Using this quantitative method, significant decreases in a number of PUFA derived from dietary polyunsaturated fatty acids is revealed (2942 ± 236 umol/L vs. 2168 ± 145* at 6 months and 2291 ± 213* at 12 months, * = $p < 0.05$) even though the ratio of PUFA to other fatty acids is not significantly changed (45.3 ± 0.91 % vs. 44.4 ± 0.99% at 6 months and 42.7 ± 2.00 % at 12 months). The control group in this study demonstrated a slight decrease in reported dietary fat intake on four day food records from 35.0 ± 6.8 % at baseline to 33.6 ± 6.8 % at 6 months, and 31.2 ± 8.2 % at 12 months. The day-to-day variation of linoleic and arachidonic acid during a 4 month period, expressed as coefficient of variation, was 5.2 % and 6.7 % respectively. There were minor changes that occurred in the composition of the fat in the diet of the control subjects resulting in minor changes in quantitative total fatty acids in the same direction as those seen in the intervention group.

6. THE RELATIONSHIP TO THE TYPE AND QUANTITY OF FAT

While decreasing dietary fat should theoretically result in a quantitative decrease in both polyunsaturated and saturated fatty acid intake, Gorbach and co-workers demonstrated that the P/S ratio (0.57) remained constant as women in the Women's Health Trial decreased total fat intake but compensated to some degree with more saturated fats derived from cheeses and dairy foods[57]. A potential difficulty in attempting to develop a biomarker for dietary fat intake is that reduction of total fat intake will result in a quantitative decrease in polyunsaturated fatty acid intake which may or may not be counterbalanced by an equivalent decrease in saturated fatty acid intake. The design of the proposed metabolic studies considers carefully the need for the biomarker to reflect type as well as quantity of fat intake.

In studies by Berlin and colleagues, twenty normal premenopausal females fed one of two sets of diets from typical U.S. foods with P/S ratios of 0.3 or 1.0[44]. All subjects were fed high fat (40% of calories) diets for four menstrual cycles followed by low fat (20% of calories) diets for the next four menstrual cycles. Blood samples were collected during mid-follicular and mid-luteal phases of the fourth menstrual cycle of each dietary period to assess interactive dietary and hormonal control of lipoprotein fluidity. LDL was significantly more fluid as determined by DPH fluorescence upon reducing fat consumption from 40% to 20% of energy for subjects eating foods with P/S = 1.0 or 0.3 as shown in the Table 3 below.

7. THE RELATIONSHIP OF LOW FAT AND HIGH FAT TO BODY WEIGHT

High fat diets are associated with increased caloric intake and weight gain compared to low fat diets[58]. In particular, postmenopausal women compared to premenopausal women have decreased lean body mass and increased upper body fat compared premenopausal women associated with a high incidence of obesity[59]. These changes are influenced by height, weight and resting metabolic rate. While these physiological changes are associated with increased dietary fat intake, they are not likely to be adequately sensitive or specific by themselves to be used as biomarkers. Nonetheless, they can be taken into consideration in the design and interpretation of the studies as potentially useful components of a pattern of biomarkers.

To date three studies have been undertaken in postmenopausal women to test the feasibility of reducing dietary fat intake for breast cancer prevention[60-62]. In each case, the intervention has been targeted solely at reducing dietary fat intake rather than at reducing body weight or body fat. Nonetheless, small but significant decreases in body weight were observed in each of these studies. A recent study by Lissner and co-workers demonstrated that caloric compensation on a low fat diet in women, (i.e. the tendency to increase dietary intake to compensate for decreased nutrient density), was only 35 percent effective[63]. These findings suggest that despite caloric compensation in postmenopausal women, a low fat diet will likely result in some small amounts of weight loss. Over the past 50 years, hundreds of epidemiological and experimental studies have been published on the association of dietary fat and breast cancer. While some authorities conclude that caloric content of the diet is more important than fat content based on animal studies[64], our research group has observed that a low fat diet in free-living women is likely to also be lower in calories[65]. Women eating a 40% fat diet ad libitum ingest 600 Cal/day more than women eating a 20% fat diet[62]. In our own controlled feeding studies[66] of a 10% fat, high fiber diet, and in our trial examining adherence to a 20% fat diet patterned on the Women's Health Trial we demonstrated small but significant average weight loss[67] in agreement with other low fat diet and breast cancer

Table 3. DPH fluorescence anisotropy at 37°C - LDL fraction

Diet (energy%, P/S)	Follicular Phase	Luteal Phase
Self-selected	0.208 ± 0.002	0.212 ± 0.002
40%, P/S 0.3	0.209 ± 0.002	0.213 ± 0.002
40%, P/S 1.0	0.204 ± 0.008	0.205 ± 0.003
20%, P/S 0.3	0.193 ± 0.003*	0.197 ± 0.004*
20%, P/S 1.0	0.190 ± 0.003*	0.196 ± 0.004*

* significantly different from self-selected and 40% diets

feasibility trials. We also found significant changes in a group of plasma lipids in our subjects after one month on the diet as shown in Table 4 below. Such a group of lipids which are likely to be modulated by nutritional intervention, both alone and in combination, may serve as biomarkers of the dietary fat intake.

8. FAT SOLUBLE VITAMINS

Although measurement of different vitamins may seem to be an easy biomarker, it presents other problems due to the complexity of their metabolism which includes regulation of their bioavailability in relation to need and intake levels. The large ratios of intra- and interindividual variance found in studies of retinol, carotenoids, and tocopherols can constitute an important source of error in studies[68]. Because antioxidant micronutrients, including vitamins E and carotenoids, might influence plasma fatty acid composition by preventing lipid peroxidation, it has been recommended that antioxidant micronutrients should be measured when polyunsaturated fatty acid metabolism is studied[69].

8.1 Carotenoids

The possibility of being a good marker of reduced fat intake seems less reasonable for vitamin A, or retinol, than for carotenoids. The carotenoids are a large family of more than 500 compounds; and at least 50 have some provitamin A activity[70].Total carotenoid storage is limited and it is apparently not regulated by a homeostatic mechanism, as is the case for retinol stored in the liver. In addition, the amount and the composition of plasma carotenoids are related to the foods consumed within a period of a few days or weeks[71,72]. Plasma levels are also influenced by their rate of intestinal absorption (30% to 60%), the efficiency of their enzymatic transformation into retinol, and their rates of clearance from the plasma[73]. The amount of fat in the diet also plays a part in the intestinal absorption, which is reduced by a very low fat intake. In a randomized trial over a 16-week period, daily supplementation of 30 mg of beta-carotene increase threefold the average plasma carotenoid levels[72].. However there is also substantial individual heterogeneity with respect to blood levels of both total and individual carotenoids and tissue levels also indicate large variation, with adipose tissue and liver seemingly the major sites of deposition[74]. In fact, total carotenoid adipose tissue concentration has been shown to vary 40-fold between individuals.

8.2 Vitamin E

Eight tocopherols and tocotrienols with vitamin E activity are known today[19]. Beta and gamma-tocopherol and alpha-tocotrienol have 40%, 8% and 20% alpha-tocopherol activity, respectively. The main dietary sources of vitamin E are vegetable oils, and to a lesser extent,

Table 4. Changes in plasma lipids and fatty acids

Plasma Levels	Baseline	Low Fat/ High Fiber:1 month
Total Cholesterol (mg/dL)	173 ± 24.5	152 ± 25*
LDL-Cholesterol (mg/dL)	104 ± 16.5	83 ± 17*
Triglycerides (mg/dL)	89 ± 25	103 ± 25
Total Linoleic Acid (umol/L)	2318 + 270	1907 + 327*

eggs, butter and cereals. Plasma levels of vitamin E depend on the dietary intake and are related to the amount and type of lipoproteins and other plasma lipids, as well as the level of antioxidants in the diet[75]. Changes in the dietary supplementation of vitamin E within the same subject causes changes in alpha-tocopherol serum levels within a few weeks [76-78]. Gamma-tocopherol has only a tenth of the Vitamin E activity of alpha-tocopherol, but it is the major tocopherol found in food and has a wide distribution in commonly used vegetable oils and products made from these oils. Subjects with higher serum alpha-tocopherol have been shown to have lower gamma-tocopherol levels and vice versa[79]. In addition, it has been found that simultaneous administration of high doses of alpha-tocopherol and beta-carotene leads to plasma levels of beta-carotene lower than those observed with administration of beta-carotene alone[72]. Although within the same subject there is good correspondence between dietary intake and plasma levels of vitamin E, the relationship is less straightforward when comparing subjects, particularly in cross-sectional studies. Total tocopherol concentrations in adipose tissue have been show to vary 11 fold[74]. A partial correlation coefficient adjusted for age, sex, total energy an cholesterol and triglyceride levels was only of the order of .30 in a carefully conducted study[78]. In a comparison among vitamin E levels and the relationships among tocopherols in human plasma, platelets, lymphocytes and red blood cells, the changes in tocopherol levels in platelets was found to follow most closely changing dietary intakes[80].

9. UNIQUE FATTY ACIDS

Trans oleic acid (C18:1t) and all the trans isomers of C18:1 are the major class of trans fatty acids found in the diet. These fatty acids are mainly formed by partial hydrogenation of unsaturated fatty acids by food manufacturers, although small amounts can be found in dairy fats. Hydrogenation of vegetable oils produces fatty acids with a variety of positional isomers of trans double bonds, mainly trans-C18:1 n-9 (elaidic acid)[80]. The major dietary sources of C18:1t are food which these hydrogenated oils, including margarine, vegetable shortening, cookies, crackers, snack foods and salad dressings/ mayonnaise[81]. The average per capita trans fatty acid consumption in the U.S. is estimated to be 12.5 g/day (with a range of 1.6 to 38.7 g/day, accounting for approximately 5% to 8% of the total fat in the diet[82]. Because trans fatty acids are found in many foods, a higher total fat intake will likely increase trans fatty acid levels. Trans fatty acids are well absorbed from the diet and can be incorporated into cell membranes or oxidized for energy. The concentrations of trans fatty acids in tissues are fairly low, usually between 0.1 and 1.0% of fatty acids in the U. S. Population. A relatively good correlation coefficient of r=.51 was found between dietary intake of trans fatty acids, assessed by a food frequency, and adipose tissue trans fatty acid concentration, with stronger correlations in the subgroup of women with stable weight over the six months before fat aspiration[30]. Of interest is the observation that the analysis of fatty acid iosmers from hydrogenated fats in human tissue showed that the double bond distribution of the triglycerides correlated closely with the pattern found in dietary hydrogenated vegetable oils, whereas the phospholipid fraction showed substantially fewer trans 18:1 isomers with evidence of selective incorporation of specific isomers with double bonds closer to the methyl end[80].

Although the fatty acids in natural dietary lipids have an even number of carbons, micro-organisms are capable of synthesizing odd number fatty acids and the rumen of ruminant animals (cows, sheep, goats) have high concentrations of the precursor of C17:0[83]. C17:0 is synthesized in the rumen and is absorbed to some extent and found in all fats derived from ruminants, including meat and dairy fat found in butter and cheese, and it is highly correlated with the total fat content of these foods.

10. CONCLUSION

A number of lipid biomarkers in both the circulation and in cells have been shown to correlate with dietary fat changes. An ideal lipid biomarker should be sensitive to dietary levels, with a low within subject variability. Given the diversity of self-selected lowfat diets, it may be that only one biomarker by itself will never be sufficient to monitor dietary compliance, and researchers are currently trying various combinations of different biomarkers to increase the predictive power of their correlations with intake. Ease of collection of samples, and the effects on subject compliance and burden, may also influence the choice of biomarkers. In addition, strict protocols for both the collection and analysis of samples need to be followed to minimize errors and bias. Finally, when measuring lipids as biomarkers of intake, it is important to understand just how the body processes dietary fat, including how the fatty acids are absorbed and metabolized and howh genetic and physiologic differences between individuals may influence results.

REFERENCES

1. U.S. Congress, Senate 1977a. Dietary Goals for the United States. Report of the Select Committee on Nutrition and Human Needs. stock No. 052-070-03913-2. U.S. Government Printing Office, Washington, D.C.
2. U.S. Department of Health and Human Services. 1988. Surgeon General's Report on Nutrition and Health. DHHS (PHS) Publication No. 88-50210. Washington, D.C.:U.S. Government Printing Office.
3. U.S. Department of Agriculture/Health and Human Services. 1990. Nutrition and Your Health: Dietary Guidelines for Americans, Bulletin No. 232. Washington, D.C.: U.S. Government Printing Office.
4. National Research Council. 1989. Diet and Health: Implications for Reducing Chronic Disease Risk. Committee on Diet and Health, Food and Nutrition Board, Commission on Life Sciences. National Academy Press, Washington, D. C.
5. Committee on the Dietary Guidelines Implementation. Food and Nutrition Board, Institute of Medicine. 1991. Improving America's Diet and Health. From Recommendations to Action. National Academy Press. Washington, D.C.
6. Wynder EL, Cohen LA. Nutritional opportunities find limitations in cancer prevention. Oncology (supplement), 1993; 7(11):13-16.
7. Ashley JM. Dietary Guidelines for Cancer Prevention and Control. Oncology (supplement), 1993; 7(11):27-31.
8. Margetts BM. (1991) Basic issues in designing and interpreting epidemiological research. In: Design Concepts in Nutritional Epidemiology, Margetts BM and Nelson M (eds), Oxford Medical Publications: Oxford.
9. Sanders TAB, Roshanni F. The influence of different types of n-3 polyunsaturated fatty acids on blood lipids and platelet function in healthy volunteers. Clin. Sci., 1983; 64:91-99.
10. Bates CJ, Thurnham DI: (1991) Biochemical markers of nutrient intake. In: Design Concepts in Nutritional Epidemiology, Margetts BM and Nelson M (eds), Oxford Medical Publications: Oxford..
11. Kris Etherton PM, Krummel D. Russell ME, Dreon D, Mackey S, Borchers J, Wood PD. The effect of diet on plasma lipids, lipoproteins, and coronary heart disease. Am. Diet. Assoc., 1988; 88:1373-1400.
12. Beynen AC. Katan MB. Why do polyunsaturated fatty acids lower serum cholesterol? Am. J. Clin. Nutr., 1985; 42, 560-563.
13. McNamara DJ: Dietary fatty acids, lipoproteins, and cardiovascular disease. Advances in Food and Nutrition Research, 1992; 36:253-351.
14. McNamara, D J, Kolb R, Parker TS, Batwin H, Samuel P, Brown CD, Ahrens EH. Heterogeneity of cholesterol homeostasis in man. Response to changes in dietary fat quality and cholesterol quantity. J. Clin. Invest. 1987; 79, 1729-1739.
15. Boyd NF, Cousins M, Beaton M, Kriukov V, Lockwood G, Tritchler D: Quantitative changes in dietary fat intake and serum cholesterol in women: results from a randomized, controlled trial. Am. J. Clin. Nutr., 1990; 52:470-476.

16. Kasim SE, Martino S, Kim P-N, Khilnani S, Boomer A, Depper J, Reading BA, Heilbrun LK. Dietary and anthropometric determinants of plasma lipoproteins during a long-term low-fat diet in healthy women. Am. J. Clin. Nutr. 1993; 57(2):146-53.
17. Katan MB, van Birgelen A, Deslypere JP, Penders M, van Staveren WA. Biological markers of dietary intake, with emphasis on fatty acids. Ann. Nutr. Metab. 1991; 35:249-252.
18. Weinberg RB, Dantzker C, Patton CS:Sensitivity of serum apolipoprotein A-IV levels to changes in dietary fat content. Gastroenterol., 1990; 98(1):17-24.
19. Davidson S, Passmore R, Brock JF, Trusswell AS: (1979) Human Nutrition and Dietetics. Churchill Livingstone, Edinburgh.
20. Dayton S, Hashimoto S, Dixon W, Pearce ML. Composition of lipids in human serum and adipose tissue during prolonged feeding of a diet high in unsaturated fat. J. Lipid. Res., 1966; 7, 103-111.
21. Berry EJ , Hirsch J, Most J, McNamara DJ, Thornton J. The relationship of dietary fat to plasma lipid levels as studied by factor analysis of adipose tissue fatty acid composition in a free-living population of middle-aged American men. Am. J. Clin. Nutr. 1986; 44:220-231.
22. Hirsch J, Farquhar JW, Ahrens EH, Peterson ML, Stoffel W. Studies of adipose tissue in man; A microtechnic for sampling and analysis. Am J. Clin Nutr., 1960; 8:499-511.
23. Beynen, AC, Hermus RJ, Hautvast JG. A mathematical relationship between the fatty acid composition of the diet and that of the adipose tissue in man. Am. J. Clin. Nutr. 1980; 33: 81-85.
24. Turpeinen O. (1973) Primary prevention of heart disease by diet. In: DWK Cotton (ed), Lipid Metabolism and Atherosclerosis, Exerpta Medica, Amsterdam, p. 49.
25. Van Staveren WJ, Durenberg P, Katan MB, Burema J, de Groot LC, Hoffmans MD. Validity of the fatty acid composition of subcutaneous fat tissue microbiopsies as an estimate of the long-term average fatty acid composition of the diet of separate individuals. Am. J. Epidemiol., 1986; 123:455-463.
26. Hunter DJ , Rimm ER, Sacks FM, Meir J, Stampfer MJ, Colditz GA, Litin B, Willett WC. Comparison of measures of fatty acid intake by subcutaneous fat aspirate, food frequency questionnaire, and diet records in a free-living population of US men. Am. J. Epidemiol. 1992; 135:418-427.
27. Riboli E., Ronnholm H, Saracci R. Biologic markers of diet. Cancer Surveys, 1987; 6, 685-718.
28. Stanford JL, King I, Kristal AR. Long-term storage of red blood cells and correlations between red cell and dietary fatty acids: results from a pilot study. Nutr. Cancer, 1991; 16:183-188.
29. Hunter DJ, Rimm EB, Sacks FM, Stampfer MJ, Colditz GA, Litin LB, Willett WC. Comparison of measures of fatty acid intake by subcutaneous fat aspirate, food freequency questionnare and diet records in free-living populations of US men. Am. J. Epidemiol. , 1992; 135:418-427
30. London SJ, Sacks FM, Caesar J, Stampfer MJ, Siguel E, and Willett WC. Fatty acid composition of subcutaneous adipose tissue and diet in postmenopausal U.S. women. Am J Clin Nutr 1991; 54:340-345.
31. Field CJ, Clandinin MT. Modulation of adipose tissue fat compositon by diet: a review. Nutr. Res., 1984; 743-755.
32. Leaf DA, Connor WE, Barstad L, Sexton G: Incorporation of dietary n-3 fatty acids into the fatty acids of human adipose tissue and plasma lipid classes. Am. J. Clin. Nutr., 1995; 62:68-73.
33. 29. Leaf DA, Connor WE, Barstad L, Sexton G: Incorporation of dietary n-3 fatty acids into the fatty acids of human adipose tissue and plasma lipid classes. Am. J. Clin. Nutr., 1995; 62:68-73.
34. Beynen AC and Katan MF. Rapid sampling and long-term storage of sub-cutaneous adipose-tissue biopsies for determination of fatty acid composition. Am. J. Clin. Nutr., 1985; 42, 317-322.
35. Smith WCS, Tavendale R, Turnstall-Pedoe H. Simplified subcutaneous biopsy for nutritional surveys. Human Nutrition: Clinical Nutriiton, 1986; 400:323-325.
36. Handelman GJ, Epstein Wl, Machlin LJ, van Kuijk F J, Dratz EA. Biopsy method for human adipose with vitamin E and lipid measurements. Lipids 1988; 23:598-604.
37. Kuo JG (ed). (1985) Phospholipids and Cellular Regulation. CRC Press, Boca Raton, Florida.
38. Field CJ, Angel A, Clendenin MT: Relationship of diet to the fatty acid composition of human adipose tissue structure and stored lipids. Am. J. Clin. Nutr., 1985; 42:1206-1220.
39. Horwitt MK, Harvey CC, Century B. Effects of dietary fats on human erythrocyte fatty acid pattern. J. Clin. Invest., 1959; 42:675-686.
40. Farquhar JW, Ahrens EH. Effects of dietary fats on human erythrocytes fatty acid pattern. J Clin Invest. 1963: 675-686.
41. Verma SP, Philippot JR, Bonnet B. Raman studies of structural rearrangements induced in human plasma lipoprotein carotenoids by malondialdehyde. Lipids, 1985; 20: 890-96.
42. Verma SP, Wallach DFH, Sakura JD. Raman analysis of the thermotropic behavior of lecithin-fatty acid systems and of their interaction with proteolipid apoprotein. Biochemistry, 1980; 19: 574-579.
43. Verma SP, Wallach DFH. Raman spectra of some saturated, unsaturated and deuterated C18 fatty acids in the HCH-deformation and CH-stretching regions. Biochim. Biophys. Acta., 1977; 486:217-227.

44. Berlin E, Judd JT, Nair PB, Jones DY, Taylor PR. Dietary fat and hormonal influences on lipoprotein fluidity and composition in premenopausal women. Atherosclerosis, 1991: 86:95-110.
45. Dougherty, R. M., C. Galli, A. Ferro-Luzzi, and J. M. Iacono, Lipid and phospholipid fatty acid composition of plasma, red blood cells, and platelets and how they are affected by dietary lipids: a study of normal subjects from Italy, Finland, and the U.S.A. Am. J. Clin. Nutr., 1987; 45, 443-455.
46. Lorenz R, Spengler U, Fischer S, Duhm J, Weber PC. Platelet function, thromboxane formation and blood pressure control during supplementation of the Western diet with cod liver oil. Circulation, 1983; 67:504-511.
47. Renaud S, Fodsey F, Dumon TF, Thevenon C, Ortchanian E, Martin JL. Influence of long-term diet modificatins on platelet functin and composition in Moselle farmers. Am. J. Clin. Nutr., 1986; 43:136-150.
48. McMurchie, E. J., B. M. Margetts, L. J. Beilin, K. D. Croft, R. Vandongen, and B. K. Armstrong. Dietary-induced changes in the fatty acid composition of human cheek cell phospholipids: Correlation with changes in the dietary polyunsaturated/saturated fat ratio. Am. J. Cain. Nutr, 1984;. 39, 965-980
49. McMurchie, E. J., J. D. Potter, T. E. Rohan, B. S. Hetze.. Human cheek cells; a non-invasive method for determining tissue lipid profiles in dietary and nutritional studies. Nutr. Rep. Int., 1984; 29, 519-526.
50. Moore RA, Oppert S,. Eaton P, Mann JL. Triglyceride fatty acids confirm a change in dietary fat. Clin. Endocrin., 1977; 143-149.
51. Yamori Y, Nara Y, Iritani N, Workman RJ, Inagami T.. Comparison of serum phospholipid fatty acids among fishing and farming Japanese populations and American inlanders. J. Nutr. Sci. Vitaminol., 1995; 31, 417-422.
52. Van Houwelingerf AC, Kester ADM, Hornstra F. Comparison between habitual intake of polyunsaturated fatty acids from fish and their concentrations in serum lipid fractions. Eur. J. Clin. Nutr., 1989; 43:11-20.
53. Lepage G, Roy CC. Direct transesterification of all lipids classes in a one step reaction. J. Lipid Res., 1986; 27:114-20.
54. Sarkkinen ES, Agren JJ, Ahola I, Ovaskainen M-L, Uusitupa MIJ. Fatty acid composition of serum cholesterol esters, and erthrocyte and platelet membranes as indicators of long-term adherence to fat-modified diets. Am. J. Clin. Nutr., 1994; 59:364-370.
55. Lopes M, Trimbo SL, Mascioli EA, Blackburn GL. Human plasma fatty acids and how they are related to dietary intake. Am. J. Clin. Nutr. 1991; 53: 628-37.
56. Chlebowski RT, Nixon DW, Blackburn GL, et al. Feasibility of using dietary fat intake reduction in adjuvant breast cancer management. Proc ASCO, 1991;10:86,.
57. Woods MN, Borbach Sl, Longscope C, Goldin Br. Low fat, high fiber diet and serum estrone sulfate in premenopausal women. Am. J. Clin. Nutr., 1989; 49:1179-1183.
58. . Lissner L, Levitsky DA, Strupp BJ, Kalkwarf HJ, Roe DA. Dietary fat and the regulation of energy intake in human subjects. Am. J. Clin. Nutr., 1987;46:886-892.
59. Bouchard C, Bray GA, Hubbard VS. Basic and clinical aspects of regional body fat distribution. Am. J. Clin. Nutr., 1990; 52:946-950.
60. Lee-Han H, Cousins M, Beaton M, McGuire V, Kriukov V, Chipman M, Boyd N. Compliance in a randomized clinical trial of dietary fat reduction in patients with breast dysplasia. Amer.J. Clin. Nut., 1988; 48:575-586.
61. Heber D, McCarthy WJ, Ashley J. Weight reduction for breast cancer prevention by restriction of dietary fat and calories: rationale, mechanisms and interventions. Nutrition, 1989; 5:149-154, 1989.
62. Insull W, Henderson MM, Prentice RL. Results of a randomized feasibility study of a low fat diet. Arch. Intern. Med. 1990;150:421-427.
63. Kendall A, Levitsky DA, Strupp BJ, Lissner L. Weight loss on a low-fat diet: consequence of the imprecision of the control of food intake in humans. Am. J. Clin. Nutr., 1991; 53:1124-1129.
64. Kritchevsky D, Weber MM, Klurfeld DM. Dietary fat vs. caloric content in initiation and promotion of 7,12 dimethyl-benzanthracene-induced mammary tumorigenesis in rats. Cancer Res., 1984; 44:3174-3177.
65. Heber D.The role of nutrition in cancer prevention and control.Oncology, 1992 ;6: 9-14.
66. Bagga D, Ashley JM, Geffrey S, Wang H-J, Barnard J, Elashoff R, and Heber D. Modulation of serum and breast ductal fluid lipids by a very low fat, high fiber diet in premenopausal women. JNCI 1994; submitted in revision.
67. Heber,D, Ashley JM, McCarthy, WJ, Solares, ME, Leaf,DA, Chang, L-J C, and Elashoff RM. Assessment of adherence to a low fat diet for breast cancer prevention. Prev. Med., 1992; 21: 218-227.
68. Tangney CC, Shekelle RB, Raymor W, Gale M, Betz EP. Intra and interindividual variation in measurements of B-carotene, retinol and tocopherols in diet and plasma. Am. J. Clin. Nutr., 1987; 45:764-769.
69. Cabre E, Periago JL, Mingorance MD, Fernandez-Banares F, Abad A, Esteve M, Gil A, Lachiaca M, Gonzalez-Huix F and MA Gassull: Factors related to the plasma fatty acid profile an healthy subjects,

with special reference to antioxidant micronutrient status, a multivariate analysis. Am. J. Clin. Nutr., 1992; 55:831-837
70. Olson, J. A. Serum levels of vitamin A and carotenoids as reflectors of nutritional status. J. Natl. Cancer Inst., 1984; 1439-1444.
71. Matthews-Roth M, Pathak MA, Fitzpatrick TB, Harbor LC, Kass EH. B-carotene as an oral photoprotective agent in erythrocyte protophorphyrin. J. Am. Med. Assoc., 1974; 228:1004-1008.
72. Willett WC, Stampfer MJ, Underwood BA, Speizer FE, B. Rosner B, Hennekens CH..Validation of a dietary questionnaire with plasma carotenoid and a-tocopherol levels. Am. J. Clin. Nutr., 183; 38, 631-639.
73. DeLuca HF. The vitamin D system in the regulation of calcium and phosphorus metabolism. Nutr. Rev., 1979; 37, 161-193.
74. Parker RS. Carotenoid and tocopherol composition of human adipose tissue.Am. J. Clin. Nutr., 1988; 47, 33-36.
75. Scott ML. Advances in understanding of vitamin E. Federation Proceedings, 1980; 39:2736-2739.
76. Farrell PM, Bieri JG. Megavitamin E supplementation in man. Am J. Clin Nutr., 1975; 28:1381-1386.
77. Garrett-Laster, Oaks M, Russell RM, Oaks RM. A lowering effect of a pharmacological dose of vitamin E on serum vitamin A in normal adults. Nutr. Res., 1981; 1:559-564.
78. Willett WC, Stampfer MJ, Underwood BA, Taylor JO,Hennekens CH. Vitamins A, E, and carotene: effects of supplementation on their plasma levels. Am. J. Clin. Nutr., 1983; 38, 559-566.
79. Behrens WA, Madre R. Alpha- and gamma-tocopherol concentrations in human serum. J. Am .Coll. Nutr., 1986; 5:91-96
80. Ohlrogge JB, Emken EA, Gulley RM. Human tissue lipids: Occurrence of fatty acid isomers from dietary hydrogenated oils. J. Lipid Res., 1981; 22, 955-960.
81. Hunter JE, Applewhite TH. Reassessment of trans fatty acid availability in the U.S. diet. Am. J. Clin. Nutr., 1991; 54:363-369.
82. Jonnalagadda SS, Mustad VA, Campagne C, Kris-Etherton PM: Margarine and plasma cholesterol-A perspective for dietitians on trans fatty acids in the diet. Perspectives in Applied Nutrition, 1995; 1(3): 9-16.
83. Nawar WW. (1985) Lipids. In: Fennema OR (ed). Food Chemistry (Second edition). New York: Marcel Dekker, pp. 139-244.

10

CHOLINE

A Nutrient That Is Involved in the Regulation of Cell Proliferation, Cell Death, and Cell Transformation

Steven H. Zeisel

Department of Nutrition
University of North Carolina at Chapel Hill
Schools of Public Health and Medicine
CB# 7400, McGavran Greenberg Building
Chapel Hill, North Carolina 27599-7400

INTRODUCTION

In human beings, liver cancer is one of the most frequent tumors, occurring mainly in areas where hepatitis B virus is endemic [1]. Dietary deficiency of choline causes hepatocellular carcinoma in the rat [2, 3]; this is the only nutritional deficiency known to cause cancer spontaneously. Though we do not believe choline deficiency is the major cause of human liver cancer, we suggest that this experimental model can be used to study how a nutrient modulates some of the molecular events that regulate hepatocyte proliferation, death and transformation.

Choline is the major methyl donor in the diet, and, in some parts of the world, diets that are at least marginally methyl-deficient are common and might play a role in promoting hepatocellular carcinoma [4]. Our model shares common features with other recognized models for hepatocellular carcinogenesis. For example, liver cell turnover increases in hepatitis virus infection, and the hepatitis virus protein Hb_x increases intracellular DAG, activates PKC, and enhances transactivation of the binding motif for transcription factor activator protein 1 (AP-1) in a PKC dependent manner [5, 6]. Similarly, in choline deficient rats, there is a large increase in the rate of hepatocyte turnover (increased proliferation and death rates), associated with the accumulation of DAG, activation of PKC, and hepatocellular carcinogenesis [2, 3].

INTER-RELATIONSHIPS BETWEEN CHOLINE, FOLATE AND METHIONINE

Choline is a precursor for the biosynthesis of acetylcholine and the choline-phospholipids (e.g., phosphatidylcholine and sphingomyelin in cell membranes) [7]. It is also the

Dietary Fats, Lipids, Hormones, and Tumorigenesis, edited by Heber and Kritchevsky
Plenum Press, New York, 1996

major dietary source for labile methyl-groups [7]. The metabolism of choline, methionine and methyl-folate are so closely inter-related that the determination of the dietary requirement for each greatly depends on the availability of the other methyl-donors [7]. The methyl group of choline can be made available, upon conversion of choline to betaine. Betaine:homocysteine methyltransferase catalyzes the methylation of homocysteine using betaine as the methyl donor. In an alternative pathway, 5-methyltetrahydrofolate: homocysteine methyltransferase regenerates methionine using a methyl group derived *de novo* from the 1-carbon pool. Methionine adenosyltransferase converts methionine to SAM (the active methylating agent for many enzymatic methylations, including the methylation of phosphatidylethanolamine to form phosphatidylcholine). Choline-deficient rats become depleted of methyl-folate, and folate deficient rats become depleted of choline pools [8-10]. Ghoshal and Farber [11] have recently argued that methyl deficiency is different from choline deficiency; both diets cause fatty liver, but pure choline deficiency uniquely induces cell death while methyl deficiency does not. Can other methyl-donors completely substitute for choline? We need to differentiate whether it is the choline moiety or methyl-groups that modulate cell proliferation, apoptosis and transformation.

CHOLINE AND CARCINOGENESIS

Choline is the only single nutrient deficiency that causes cancer in the absence of the addition of any known carcinogen [12-14]. Choline deficiency also makes it much more likely that administration of a carcinogen, such as aflatoxin, will result in carcinogenesis [14]. The earliest changes in liver occur within days to weeks after a rat begins a choline deficient diet. These changes include fatty liver, a 30-fold increase in hepatocyte DNA synthesis [15] and matching increase in hepatocyte death rates [16], and depletion of choline-compounds and *S*-adenosylmethionine (SAM) concentrations [7]. Choline deficiency causes preneoplastic foci to appear in liver at approximately 6 months, and hepatocellular carcinoma at 12 months [3, 13]. At 52 weeks on semisynthetic experimental diets, 15% of the choline deficient rats developed hepatocellular carcinoma and 8% had hepatocellular adenoma [3]. The tumors were well differentiated hepatocellular carcinomas of the trabecular and adenomatous type previously described [17,18]. 77% of the deficient animals had abnormal hepatic foci that expressed

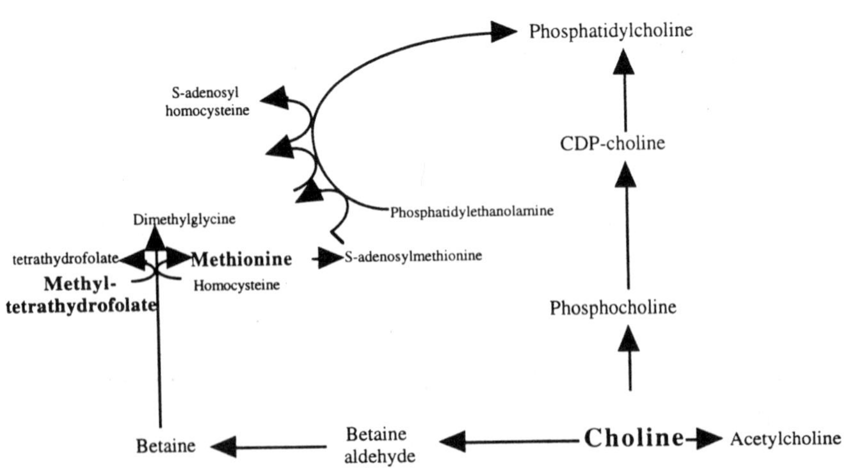

Figure 1. Pathways for Choline Metabolism.

γ-glutamyltranspeptidase (5681 ± 2308 foci/liver). A diet containing choline completely prevents the development of cancer [3, 13].

The mechanism for the cancer-promoting effect of choline deficiency is not completely elucidated, but several hypotheses have been suggested. Alterations in the genome that occur in choline deficiency could contribute to the events leading to carcinogenesis. Enhanced free radical generation, with associated premutagenic (8-oxydeoxyguanosine (8-OHdg) residues) and mutagenic alterations of DNA, occurs in liver of rats fed a choline deficient diet [19, 20]. The extent of free radical damage to liver DNA is diminished when saturated fat is removed from the choline deficient diet [21]. Hypomethylation of DNA [22-24] and abnormalities in DNA repair [25] occur in choline deficient liver. Increased liver cell turnover will increase the chance that DNA alterations occur, and thus, may be a critical component of choline deficiency carcinogenesis [26, 27].

Choline deficiency is associated with a 4-fold increase in hepatic plasma membrane DAG (see figure 2) with activation and over expression of PKC_α and PKC_δ [2, 3]. Plasma membrane DAG was higher in deficient animals through 6 weeks, but thereafter did not differ as controls exhibited an age-related increase which has not been previously appreciated (see figure 2) [3]. Choline deficient animals had 10 times more DAG in intracellular fat droplets at 52 weeks than did controls. The elevation in DAG persists for many months, and the increase in concentration is of the magnitude needed to activate PKC. There are multiple isoforms of PKC encoded for by distinct genes [28]. Some isoforms require calcium and DAG (α, β_1, β_2, γ), some only require DAG (δ, ε), and some require neither for activation (ζ, λ)[29, 30]. Chronic elevation of DAG can induce DNA synthesis, cause cell proliferation, and contribute to cell transformation, presumably via abnormal PKC signaling [31, 32]. Free fatty acids accumulate during choline deficiency (see figure 3) [3]; these can act synergistically with DAG to activate PKC [33]. In choline deficiency the activation of PKC persists for many months (see figure 4) [2, 3]. This is remarkable because PKC normally is activated only for a short time after a receptor has been stimulated [34]. Although, prolonged activation by phorbol esters leads to proteolysis of PKC [35], prolonged activation of PKC by DAG is not always associated with such down-regulation [34, 36].

Chronically increased DAG concentration with subsequent activation of PKC may perturb signal transduction and result in carcinogenesis. The phorbol esters, potent mitogens and tumor promoters, are analogs of DAG that activate PKC [35]. Several lines of evidence indicate that cancers might develop secondary to abnormalities in PKC-mediated signal transduction [28]. Oncogene expression is often associated with alterations in DAG and PKC mediated pathways [28]. Over expression of PKC_ε in liver cells caused them to transform [37], while over expression of $PKC_{\beta 1}$ caused changes in growth control [38, 39]. The targets for phosphorylation by PKC have not been completely described, but they include receptors for

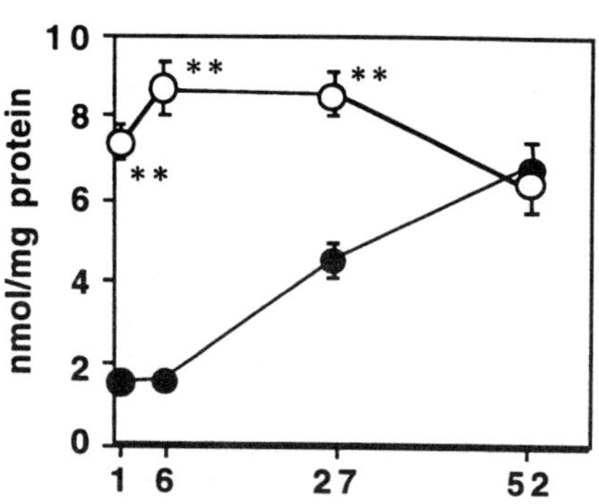

Figure 2. Diradylglycerol concentrations in rat liver. Rats were fed defined diets containing choline (control) or devoid of choline (deficient) for 52 weeks. Liver was collected and 1,2-*sn*-diradylglycerol was extracted and measured using a radioenzymatic assay. Data are expressed as mean ± SEM (n=6-14/group). Statistical differences were determined using unpaired t-Test. **= p< 0.01 different from control at same time point. From reference [3] with permission. J = control; E = deficient.

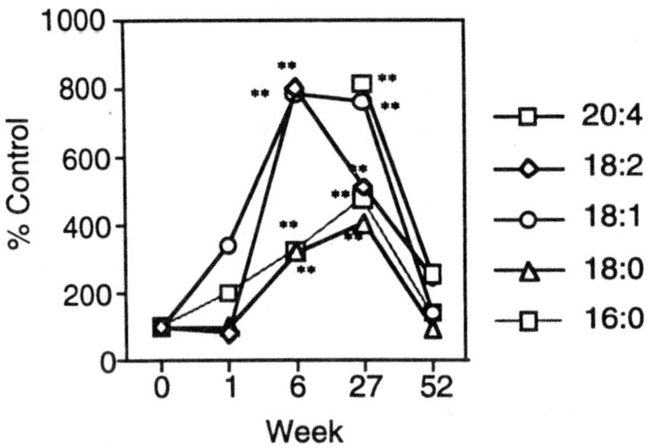

Figure 3. Free fatty acid concentrations in liver. Rats were fed as in fig. 2 legend. Free fatty acids were extracted from liver and measured using gas chromatography. Data are expressed as % control (mean ± SEM; n=5/group), where control is 100. **p< 0.01 different from control. From reference [3].

insulin and epidermal growth factor (EGF_r), and many proteins involved in control of gene expression, cell division and differentiation [28]. These include DNA methyltransferase [40], and ADP-ribosyltransferase [41].

Components from each of the above hypotheses can be combined to develop an overall hypothesis. Changes in cell signaling and gene expression may be responsible for priming of the hepatocytes (moving them from G_0 to G_1), enhanced DNA synthesis and cell death. Defects in DNA (endogenously initiated, hypomethylated or damaged DNA) will be replicated during repeated rounds of cell death and compensatory cell proliferation. This may select for the survival and growth of preneoplastic hepatocytes [42].

CHOLINE DEFICIENCY CAUSES CELL PROLIFERATION

Carcinogenesis requires enhanced cell proliferation. Within 1 day after rats begin on a methyl-deficient diet, DNA synthesis (as measured by incorporation of bromodeoxyuridine) increases by as much as 30-fold [15]. Hepatocytes are normally not dividing (38). The choline deficiency-mediated molecular mechanisms which change quiescent hepatocytes to cells competent for proliferation might involve Fos/Jun and Myc. Fos and Jun form dimeric complexes that bind to AP-1 and regulate gene expression and entry of hepatocytes into the cell cycle. Passage from G_0 to G_1 stages of cell cycle is characterized by an increase in c-Fos transcripts [43]. c-Myc expression also pushes cells to enter the cell cycle [44]. c-Fos and c-Myc are hypomethylated and over-expressed in both tumor and non-tumor tissue from choline (or choline and methyl) deficient rat liver [4, 21, 24, 44, 45].

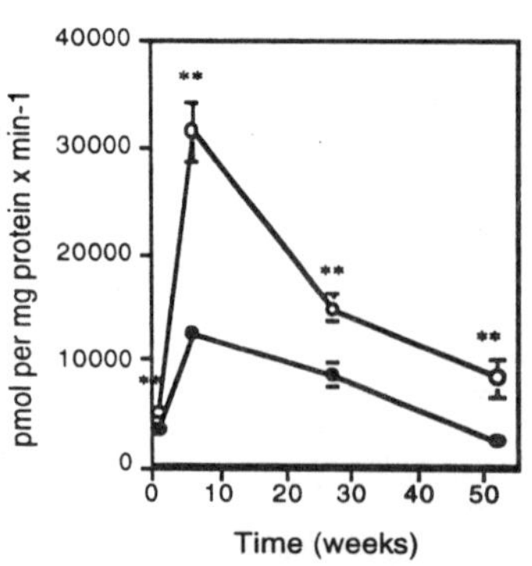

Figure 4. Plasma membrane PKC activity in rat liver. Rats were fed as in figure 1. PKC activity was measured using a radioenzymatic assay. Data are expressed as mean ± SEM (n=6/group). Statistical differences were determined using 2-way ANOVA. **= p< 0.01 different from control at same time point. From reference [3]. J = control; E = deficient.

Extracellular factors that stimulate hepatocyte proliferation include epidermal growth factor (EGF), transforming growth factor α (TGF_{α}) and hepatocyte growth factor (HGF). EGF and TGF_{α} share a receptor (EGF_r) which has intrinsic tyrosine kinase activity. HGF binds to the Met protein tyrosine kinase receptor[46]. TGF_{α} and HGF are synthesized by hepatocytes, EGF is not[43]. HGF and TGF_{α} mRNA increase by 4 weeks of choline deficiency in rat liver[47].

It is possible that the increased cell proliferation that occurs in choline deficient liver is caused by activated PKC. Sustained activation of PKC causes mitogenesis[28]. Activation of PKC, in keratinocytes, increases TGF_{α} mRNA and protein levels[48]. Active TGF_{α}(soluble) is derived from an inactive membrane-bound proTGF_{α}. Activated PKC enhances cleavage of proTGF_{α}, thereby increasing TGF_{α}-stimulated mitogenesis[49, 50].

Cell proliferation could also occur because of lack of suppression by p53 or Rb. Mutations in p53 occur at high frequency in tumors and in non-tumor liver from choline deficient rats[21]. These mutations occur in different codons, or bases within codons[21] (at one time it was thought that choline deficiency sensitized rats to low levels of aflatoxin in the diet[14]; in choline deficiency there is no p53 mutation hot spot at codon 249 that is typical of aflatoxin B_1 exposure).

CHOLINE DEFICIENCY AND CELL DEATH

Though it has been proposed that choline deficiency kills liver cells because they accumulate massive amounts of fat and rupture their plasma membranes[21], we have observed that choline deficient hepatocytes begin to die long before lipid accumulation threatens membrane integrity. Choline deficiency, in both rats and in cultured cells, triggers apoptosis[51], a physiological form of cell death[52, 53]. The nuclear changes occurring in apoptosis are the result of endonuclease activity that cleaves transcriptionally active nuclear DNA[52], but not mitochondrial DNA[54], into internucleosomal fragments. These DNA strand breaks are induced before actual apoptosis is detectable[55]. In contrast to necrosis, apoptosis is an energy-dependent process, requiring functioning mitochondria. Apoptosis can be induced by a variety of agents, including growth factors (e.g., transforming growth factor ß1 ($TGF_{ß1}$))[56], withdrawal of growth factors (e.g. TGF_{α}, EGF, spermine)[57, 58], a rise in intracellular calcium[59], a rise in intracellular ceramide concentration[60] and by tumor suppressor gene products (e.g., p53 or Rb)[61]. Several of these mechanisms might contribute to apoptosis in choline deficiency.

Withdrawal of EGF/TGF_{α} stimulation triggers apoptosis[58]. The EGF_r[62] is activated by autophosphorylation and inhibited by PKC mediated phosphorylation which induces the internalization of the EGF_r[63]. Within 1 week of beginning rats on a choline-deficient (or methyl-deficient) diet, hepatic EGF_rmRNA expression is diminished significantly compared to controls[4, 47, 64]. $TGF_{\beta 1}$ inhibits DNA synthesis[65] and induces apoptosis in hepatocytes[66, 67]. Activators of PKC enhance the synthesis of $TGF_{\beta 1}$ mRNA[68]. By 1 week after rats begin a choline deficient diet, $TGF_{\beta 1}$ mRNA synthesis in liver is increased, and this elevation persists for 6 weeks[47].

Ceramide and sphingosine modulate cell proliferation and apoptosis. We found that rats fed a choline deficient diet for 1 week had significantly higher concentrations of ceramide in their hepatic plasma membrane than did controls (107 ± 4 vs. 45 ± 3 nmol/mg protein; $p<0.01$). Sphingosine concentrations were 2-3 times higher in the deficient livers at 1, 6, 27 and 52 weeks compared to control. Plasma membrane sphingosine levels were not significantly different from control. Ceramide, a product of sphingomyelin hydrolysis, induces apoptosis[60]. For this reason, we determined whether livers from choline deficient rats had increased rates of apoptosis. Choline deficiency induced a remarkable increase in apoptosis in liver (see figure 5).

Abnormal methylation patterns are believed to disrupt normal regulation of DNA replication and repair, and thus might trigger apoptosis. Hypomethylation of DNA occurs in nuclei from choline deficient liver due to depletion of SAM [15].

A CELL CULTURE MODEL FOR CHOLINE DEFICIENCY

An ideal model system to reproduce the effects of choline-methyl-deficiency in liver would use a cell line that is derived from normal liver, has liver-like expression of liver-specific proteins and grows in defined medium without added serum. We use the SV40 large T antigen immortalized CWSV1 cell line (provided as a gift by Dr. Harriet Isom at Pennsylvania State University) which was created from normal male rat (Fischer 344) hepatocytes [69-71]. Microscopically, CWSV1 hepatocytes appear similar to mature, differentiated hepatocytes. They secrete the liver-specific protein albumin, have albumin mRNA levels identical to whole liver, and express other liver-specific proteins including transferrin, the 3rd component of complement (C3), hemopexin, glucose-6-phosphatase, tyrosine aminotransferase activity which is inducible by glucocorticoids [70], α1-antitrypsin, and phosphoenolpyruvate carboxykinase [71]. CWSV1 do not express α-fetoprotein (AFP) or AFP DNA [70]; AFP is a plasma protein expressed in fetal liver, regenerating liver and in many liver tumor cells and is associated with loss of differentiation. CWSV1 cells (passage 20-50) are not tumorigenic in syngeneic newborn rats [71].

We have shown that these cells, when grown in a serum free choline-deficient medium, recapitulate many of the biochemical changes we observed in choline deficient rat liver (Table 1).

We observed that choline-deficient CWSV1 hepatocytes synthesized DNA more rapidly than controls but did not increase in cell number (DNA content) and concluded that counterbalancing cell death must have occurred. Choline deficiency did indeed induce apoptosis in our cells (see figure 6). Cells undergoing DNA fragmentation retained the ability to exclude trypan blue, and showed an increased ability to convert

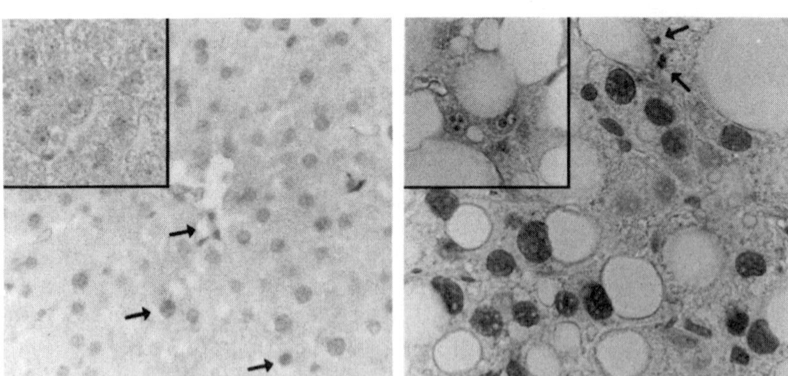

Figure 5. Incorporation of digoxygenin-11dUTP into hepatocytes to show the effects of choline deficiency on DNA fragmentation *in vivo*. Liver section from a control rat (**A**) and from a choline deficient rat (**B**) after 6 weeks on diet. DNA fragmentation was observed rarely in control liver (small arrows), whereas a high percentage of choline deficient hepatocytes had fragmented DNA (stained). Note the apoptotic cell in panel B (large arrow). Inserts are histochemical controls. x 600. From reference [51].

Table 1. Choline deficient CWSV1 hepatocytes accumulated TAG, DAG and ceramide and depleted choline pools

	nmol/mg DNA				
	Triacylglycerol	DAG	Choline	Phospho-choline	Ceramide
Control	132 ± 17	4.1 ± 0.2	11.5 ± 0.7	141 ± 3	21 ± 1
Choline deficient	418 ± 21**	5.8 ± 0.6*	1.1 ± 0.1**	8 ± 2**	63 ± 4**

Cells were grown in Eagle's MEM medium with growth factors and 21 μM choline. Choline deficient medium contained only 2 μM choline. Cells were grown for 4 days in experimental medium to approximately 70% confluence and then analyzed.

a tetrazolium salt (MTT) to an insoluble formazan product. The initiation of DNA fragmentation was followed by the progressive, dose-dependent accumulation of end-stage apoptotic cells exhibiting nuclear chromatin condensation, pyknosis, and fragmentation of the nucleus found within intact cytoplasm. Choline deficient medium not only reduced the growth fraction of CWSV-1 hepatocytes, but also inhibited the ability of cells to transit the cell cycle.

CWSV1 HEPATOCYTES WEANED FROM CHOLINE IN MEDIUM

We have been able to gradually acclimatize CWSV1 hepatocytes to a choline deficient medium by reducing choline in the medium in increments over several months. These "weaned" cells microscopically appear similar to cells grown for the same number of passages in 70 μM choline medium, though they appear to have lost contact inhibition of growth. The weaned cells increase their DNA mass only slightly more slowly than do the control cells. We assessed apoptosis using the end labeling technique in cells weaned to 8 μM choline and found that 1.2 ± 0.2% of these cells were apoptotic, compared to 0.5 ± 0.2% apoptosis in 70 μM choline cells, and 3.8 ± 0.8% apoptosis in 70 μM cells acutely changed to 8 μM choline for 4 days. Thus, these weaned cells have adapted so that choline deficiency is not as effective an inducer of apoptosis.

The weaned cells are more efficient at forming colonies in soft agar, an indication of cell transformation. Cells were gradually weaned so that they grew in diminished choline concentration over several months. Selected cells were plated in soft agar containing 70 μM choline, and colony forming efficiency was calculated by counting colonies with more than 20 cells and dividing by the number of cells plated. We found that as

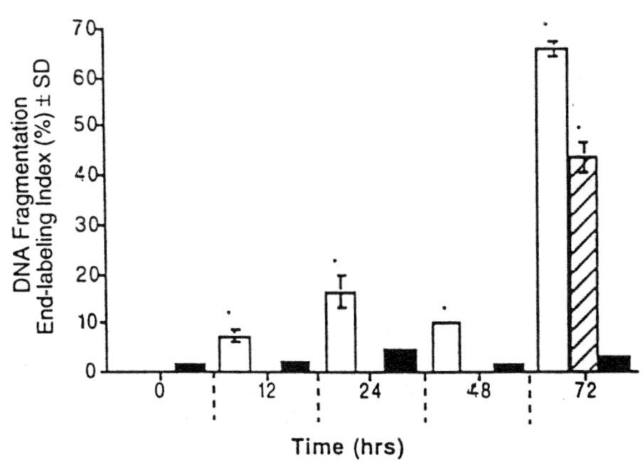

Figure 6. Effect of choline deficiency on DNA fragmentation. CWSV-1 rat hepatocytes were plated in 70 μM choline sufficient serum-free medium (CS) for 4 days and then cultured in CS or choline deficient (5 μM or 0 μM choline) medium for 3 days. The height of the bars shows the percentage of cells with nuclear incorporation of digoxigenin-11-dUTP. Open bars = 0 μM choline, cross-hatched bars = 5 μM choline, black bars = 70 μM choline. Data are expressed as mean ± SD; n = 3/treatment (t-test: * = p < .05). From reference[51].

cells adjust to lower choline content of medium they become more efficient at forming colonies in soft agar (Table 2). It did not matter whether weaned cells were placed in agar containing 70 μM choline or the amount of choline they were accustomed to, colony forming efficiency was higher in the weaned cells.

SUMMARY

Choline deficiency causes hepatocyte proliferation, apoptosis and transformation. Thus, it is an excellent model in which to study the molecular mechanisms underlying these processes. Several interesting questions can be addressed. What is the first event that begins the cells on the path towards transformation? Is it triggered by some autocrine factor produced in choline depleted cells? Does it involve alteration of DNA structure with subsequent apoptosis, compensatory cell proliferation, and enhanced survival of preneoplastic cells? Is there a specific choline deficiency signal which triggers apoptosis, with subsequent compensatory cell proliferation in a methyl-deficient environment causing hypomethylation of DNA? Does this result in abnormal transcription of genes with resulting transformation? Or is the activation of PKC the first event? PKC-mediated cell proliferation might then be balanced by down regulation of growth factor response, withdrawal of which causes apoptosis. The ensuing high rate of cell turnover might result in the survival and replication of preneoplastic cells. Multiple alternative variations of these questions exist. Whatever the critical first event is, our models also allow us to ask about molecular differences between cells that pass through these early events and those that do not. At first glance, choline deficiency may seem to be an artificial situation that might rarely occur in nature. However, the answers to some of the above questions will help us to understand how changes in gene expression and the signaling pathways that are fundamental for many cell functions, might be involved in liver cell proliferation, death and transformation.

ACKNOWLEDGMENTS

The work reviewed in this manuscript was performed by a team of investigators under Dr. Zeisel's direction, including Drs. Kerry-Ann da Costa, Craig Albright and Rong Liu, as well as Ms. Mei-Heng Mar. This work was supported by a grant from the American Institute for Cancer Research.

Table 2. Colony formation in soft agar

Cells weaned to [choline]	Cell passage	Colony forming efficiency %
70 μM	29	2.9 ± 0.1
70 μM	31	2.5 ± 0.1
70 μM	34	2.2 ± 0.1
70 μM	35	2.3 ± 0.1
20 μM	29	2.7 ± 0.1
12 μM	31	7.6 ± 0.1**
8 μM	34	15.7 ± 0.8 **
5 μM	35	34.8 ± 0.4 **

Cells were plated in soft agar containing 70 μM choline and incubated for 10 days. Colonies containing more than 20 cells were counted and this number was divided by the number of cells plated. **= $p < 0.01$ different from 70 μM at same passage by t-test.

REFERENCES

1. Shafritz, D. A. (1982). Hepatitis B virus DNA molecules in the liver of HBs Ag carriers. Mechanistic considerations in the pathogenesis of hepatocellular carcinoma. *Hepatology*. 2, 359-419.
2. daCosta, K., Cochary, E. F., Blusztajn, J. K., Garner, S. C. and Zeisel, S. H. (1993). Accumulation of 1,2-*sn*-diradylglycerol with increased membrane-associated protein kinase C may be the mechanism for spontaneous hepatocarcinogenesis in choline deficient rats. *J. Biol. Chem.* 268, 2100-2105.
3. da Costa, K.-A., Garner, S. C., Chang, J. and Zeisel, S. H. (1995). Effects of prolonged (1 year) choline deficiency and subsequent refeeding of choline on 1,2,-sn-diradylglycerol, fatty acids and protein kinase C in rat liver. *Carcinogenesis*. 16, 327-334.
4. Wainfan, E. and Poirier, L. A. (1992). Methyl groups in carcinogenesis: effects on DNA methylation and gene expression. *Cancer Res.* 52, 2071s-2077s.
5. Luber, B., Lauer, U., Weiss, L., Hohne, M., Hofschneider, P. H. and Kekule, A. S. (1993). The hepatitis B virus transactivator HBx causes elevation of diacylglycerol and activation of protein kinase C. *Res. Virol.* 144, 311-21.
6. Kekule, A. S., Lauer, U., Weiss, L., Luber, B. and Hofschneider, P. H. (1993). Hepatitis B virus transactivator HBx uses a tumour promoter signalling pathway. *Nature*. 361, 742-5.
7. Zeisel, S. H. and Blusztajn, J. K. (1994). Choline and human nutrition. *Ann. Rev. Nutr.* 14, 269-296.
8. Kim, Y.-I., Miller, J. W., da Costa, K.-A., Nadeau, M., Smith, D., Selhub, J., Zeisel, S. H. and Mason, J. B. (1995). Folate deficiency causes secondary depletion of choline and phosphocholine in liver. *J. Nutr.* 124, 2197-2203.
9. Varela-Moreiras, G., Selhub, J., da Costa, K. and Zeisel, S. H. (1992). Effect of chronic choline deficiency in rats on liver folate content and distribution. *J. Nutr. Biochem.* 3, 519-522.
10. Selhub, J., Seyoum, E., Pomfret, E. A. and Zeisel, S. H. (1991). Effects of choline deficiency and methotrexate treatment upon liver folate content and distribution. *Cancer Res.* 51, 16-21.
11. Ghoshal, A. K. and Farber, E. (1993). Choline deficiency, lipotrope deficiency and the development of liver disease including liver cancer: a new perspective. *Lab. Invest.* 68, 255-260.
12. Mikol, Y. B., Hoover, K. L., Creasia, D. and Poirier, L. A. (1983). Hepatocarcinogenesis in rats fed methyl-deficient, amino acid-defined diets. *Carcinogenesis*. 4, 1619-1629.
13. Ghoshal, A. K. and Farber, E. (1984). The induction of liver cancer by dietary deficiency of choline and methionine without added carcinogens. *Carcinogenesis*. 5, 1367-1370.
14. Newberne, P. M. and Rogers, A. E. (1986). Labile methyl groups and the promotion of cancer. *Ann. Rev. Nutr.* 6, 407-432.
15. Christman, J. K., Chen, M.-L., Sheiknejad, G., Dizik, M., Abileah, S. and Wainfan, E. (1993). Methyl deficiency, DNA methylation and cancer: Studies on the reversibility of the effects of a lipotrope-deficient diet. *J. Nutr. Biochem.* 4, 672-680.
16. Chandar, N., Amenta, J., Kandala, J. C. and Lombardi, B. (1987). Liver cell turnover in rats fed a choline-devoid diet. *Carcinogenesis.* 8, 669-73.
17. Farber, E. (1976). The pathology of experimental liver cell cancer. In *Liver Cell Cancer*. (H. M. Cameron, D. A. Linsell and G. P. Warwick ed.), Elsevier. Amsterdam. p. 243-277
18. Maronpot, R. R., Montgomery Jr., C. A., Boorman, G. A. and McConnell, E. E. (1986). National toxicology program nomenclature for hepatoproliferative lesions of rats. *Current Topics Toxicol. Pathol.* 14, 263-273.
19. Nakae, D., Yoshiji, H., Maruyama, H., Kinugasa, T., Denda, A. and Konishi, Y. (1990). Production of both 8-hydroxyguanosine in liver DNA and γ-glutamyltransferase-positive hepatocellular lesions in rats given a choline-deficient, L-amino acid-defined diet. *Jpn. J. Cancer Res.* 81, 1081-1084.
20. Banni, S., Evans, R. W., Salgo, M. G., Corongiu, F. P. and Lombardi, B. (1990). Conjugated diene and *trans* fatty acids in a choline-devoid diet hepatocarcinogenic in the rat. *Carcinogenesis*. 11, 2047-2051.
21. Lombardi, B. and Smith, M. L. (1994). Tumorigenesis, protooncogene activation, and other gene abnormalities in methyl deficiency. *J. Nutr. Biochem.* 5, 2-9.
22. Locker, J., Reddy, T. V. and Lombardi, B. (1986). DNA methylation and hepatocarcinogenesis in rats fed a choline devoid diet. *Carcinogenesis*. 7, 1309-1312.
23. Wainfan, E., Dizik, M., Stender, M. and Christman, J. (1989). Rapid appearance of hypomethylated DNA in livers of rats fed cancer promoting, methyl-deficient diets. *Cancer Res.* 49, 4094-4097.
24. Dizik, M., Christman, J. K. and Wainfan, E. (1991). Alterations in expression and methylation of specific genes in livers of rats fed a cancer promoting methyl-deficient diet. *Carcinogenesis*. 12, 1307-1312.
25. Li, D. H., Xu, D. C., Chandar, N., Lombardi, B. and Randerath, K. (1990). Persistent reduction of indigenous DNA modification (I-compound) levels in liver DNA from male Fischer rats fed choline-devoid diet and in DNA of resulting neoplasms. *Cancer Res.* 50, 7577-80.

26. Chandar, N. and Lombardi, B. (1988). Liver cell proliferation and incidence of hepatocellular carcinomas in rats fed consecutively a choline-devoid and a choline-supplemented diet. *Carcinogenesis*. 9, 259-263.
27. Abanobi, S. E., Lombardi, B. and Shinozuka, H. (1982). Stimulation of DNA synthesis and cell proliferation in the liver of rats fed a choline-devoid diet and their suppression by phenobarbital. *Cancer Res.* 42, 412-5.
28. Zeisel, S. H. (1993). Choline phospholipids: signal transduction and carcinogenesis. *FASEB J.* 7, 551-557.
29. Weinstein, I. B. (1990). The role of protein kinase C in growth control and the concept of carcinogenesis as a progressive disorder in signal transduction. *Adv. Second Messenger Phosphoprotein Res.* 24, 307-316.
30. Nishizuka, Y. (1992). Intracellular signaling by hydrolysis of phospholipids and activation of protein kinase C. *Science*. 258, 607-614.
31. Kato, M., Kawai, S. and Takenawa, T. (1989). Defect in phorbol acetate-induced translocation of diacylglycerol kinase in *erb*B-transformed fibroblast cells. *FEBS Lett.* 247, 247-250.
32. Johansen, T., Bjorkoy, G., Overvatn, A., Diaz-Meco, M. T., Traavik, T. and Moscat, J. (1994). NIH 3T3 cells stably transfected with the gene encoding phosphatidylcholine-hydrolyzing phospholipase C from *Bacillus cereus* acquire a transformed phenotype. *Mol. Cell. Biol.* 14, 646-654.
33. Lester, D. S., Collin, C., Etcheberrigaray, R. and Alkon, D. L. (1991). Arachidonic acid and diacylglycerol act synergistically to activate protein kinase C *in vitro* **and** *in vivo*. *Biochem. Biophys. Res. Comm.* 179, 1522-1528.
34. Price, B. D., Morris, J. D., Marshall, C. J. and Hall, A. (1989). Stimulation of phosphatidylcholine hydrolysis, diacylglycerol release, and arachidonic acid production by oncogenic *ras* is a consequence of protein kinase C activation. *J. Biol. Chem.* 264, 16638-16643.
35. Nishizuka, Y. (1986). Studies and perspectives of protein kinase C. *Science*. 233, 305-312.
36. Diaz-Laviada, I., Larrodera, P., Diaz-Meco, M., Cornet, M. E., Guddal, P. H., Johansen, T. and Moscat, J. (1990). Evidence for a role of phosphatidylcholine-hydrolysing phospholipase C in the regulation of protein kinase C by *ras* and *src* oncogenes. *Embo J.* 9, 3907-3912.
37. Cacace, A. M., Guadagno, S. N., Krauss, R. S., Fabbro, D. and Weinstein, I. B. (1993). The epsilon isoform of protein kinase C is an oncogene when overexpressed in rat fibroblasts. *Oncogene*. 8, 2095-2104.
38. Housey, G. M., Johnson, M. D., Hsiao, W. L. W., O'Brian, C. A., Murphy, J. P., Kirschmeier, P. and Weinstein, I. B. (1988). Overproduction of protein kinase C causes disordered growth control in rat fibroblasts. *Cell*. 52, 343-354.
39. Krauss, R., Housey, G., Johnson, M. and Weinstein, I. B. (1989). Disturbances in growth control and gene expression in a C3H/10T1/2 cell line that stably overproduces protein kinase C. *Oncogene*. 4, 991-998.
40. dePaoli-Roach, A., Roach, P. J., Zucker, K. E. and Smith, S. S. (1986). Selective phosphorylation of human DNA methyltransferase by protein kinase C. *FEBS LETTERS*. 197, 149-153.
41. Bauer, P. I., Farkas, G., Buday, L., Mikala, G., Meszaros, G., Kun, E. and Farago, A. (1992). Inhibition of DNA binding by the phosphorylation of poly ADP-ribose polymerase protein catalysed by protein kinase C. *Biochem Biophys Res Comm.* 187, 730-736.
42. Harris, C. C. and Sun, T. (1984). Multifactoral etiology of human liver cancer. *Carcinogenesis*. 5, 697-701.
43. Fausto, N. and Mead, J. E. (1989). Regulation of liver growth: protooncogenes and transforming growth factors. *Lab. Invest.* 60, 4-13.
44. Hsieh, L. L., Wainfan, E., Hoshina, S., Dizik, M. and Weinstein, I. B. (1989). Altered expression of retrovirus-like sequences and cellular oncogenes in mice fed methyl-deficient diets. *Cancer Res.* 49, 3795-3799.
45. Chandar, N., Lombardi, B. and Locker, J. (1989). c-myc gene amplification during hepatocarcinogenesis by a choline-devoid diet. *Proc. Natl. Acad. Sci. USA*. 86, 2703-2707.
46. Boccaccio, C., Gaudino, G., Gambarotta, G., Galimi, F. and Comoglio, P. M. (1994). Hepatocyte growth factor (HGF) receptor expression is inducible and is part of the delayed-early response to HGF. *J. Biol. Chem.* 269, 12846-51.
47. Shinozuka, H., Masuhara, M., Kubo, Y. and Katyal, S. L. (1993). Growth factor and receptor modulations in rat liver by choline-methionine deficiency. *J. Nutr. Biochem.* 4, 610-617.
48. Pittelkow, M. R., Lindquist, P. B., Abraham, R. T., Graves-Deal, R., Derynck, R. and Coffey, R. J. J. (1989). Induction of transforming growth factor-alpha expression in human keratinocytes by phorbol esters. *J. Biol. Chem.* 264, 5164-71.
49. Bosenberg, M. W., Pandiella, A. and Massague, J. (1993). Activated release of membrane-anchored TGF-alpha in the absence of cytosol. *J. Cell. Biol.* 122, 95-101.
50. Massague, J. and Pandiella, A. (1993). Membrane-anchored growth factors. *Annu Rev Biochem.* 62, 515-41.
51. Albright, C. D., Bethea, T. C., da Costa, K.-A. and Zeisel, S. H. (1995). Choline deficiency induces apoptosis in SV40-immortalized CWSV-1 rat hepatocytes in culture. *submitted*.

52. Arends, M. J., Morris, R. G. and Wyllie, A. H. (1990). Apoptosis. The role of the endonuclease. *Am. J. Pathol.* 136, 593-608.
53. Wyllie, A. H. (1987). Cell death. *Int. Rev. Cytol.* 17 (Suppl), 755-785.
54. Murgia, M., Pizzo, P., Sandona, D., Zanovello, P., Rizzuto, R. and Virgilio, F. D. (1992). Mitochondrial DNA is not fragmented during apoptosis. *J. Biol. Chem.* 267, 10939-10941.
55. Nosseri, C., Coppola, S. and Ghibelli, L. (1994). Possible involvement of poly(ADP-ribosyl) polymerase in triggering stress-induced apoptosis. *Exp Cell Res.* 212, 367-73.
56. Rotello, R. J., Lieberman, R. C., Purchio, A. F. and Gerschenson, L. E. (1991). Coordinated regulation of apoptosis and cell proliferation by transforming growth factor beta 1 in cultured uterine epithelial cells. *Proc. Natl. Acad. Sci. U.S.A.* 88, 3412-5.
57. Brune, B., Hartzell, P., Nicotera, P. and Orrenius, S. (1991). Spermine prevents endonuclease activation and apoptosis in thymocytes. *Exp Cell Res.* 195, 323-9.
58. Schwartzman, R. A. and Cidlowski, J. A. (1993). Apoptosis: The biochemistry and molecular biology of programmed cell death. *Endoc. Rev.* 14, 133-151.
59. Martikainen, P., Kyprianou, N., Tucker, R. W. and Isaacs, J. T. (1991). Programmed death of nonproliferating androgen-independent prostatic cancer cells. *Cancer Res.* 51, 4693-700.
60. Obeid, L. M., Linardic, C. M., Karolak, L. A. and Hannun, Y. A. (1993). Programmed cell death induced by ceramide. *Science.* 259, 1769-1771.
61. Yonish-Rouach, E., Resnitzky, D., Lotem, J., Sachs, L., Kimchi, A. and Oren, M. (1991). Wild-type p53 induces apoptosis of myeloid leukaemic cells that is inhibited by interleukin-6. *Nature.* 352, 345-7.
62. McCune, B. K., Prokop, C. A. and Earp, H. S. (1990). Transient epidermal growth factor (EGF)-dependent suppression of EGF receptor autophophorylation during internalization. *J. Biol. Chem.* 265, 9715-9721.
63. Beguinot, L., Hanover, J. A., Ito, S., Richert, N. D., Willingham, M. C. and Pastan, I. (1985). Phorbol Esters Induce Internalization Without Degradation of Unoccupied Epidermal Growth Factor Receptors. *Proc. Natl. Acad. Sci. USA.* 82, 2774-2778.
64. Gupta, C., Hattori, A., Betschard, J. M., Virji, M. A. and Shinozuka, H. (1988). Modulation of epidermal growth factors in rat hepatocytes by two liver tumor-promoting regiments, a choline-deficient and a phenolbarbital diet. *Cancer Res.* 48, 1162-1165.
65. Russell, W. E. (1988). Transforming growth factor beta (TGF-beta) inhibits hepatocyte DNA synthesis independently of EGF binding and EGF receptor autophosphorylation. *J Cell. Physiol.* 135, 253-61.
66. Oberhammer, F. A., Pavelka, M., Sharma, S., Tiefenbacher, R., Purchio, A. F., Bursch, W. and Schulte-Hermann, R. (1992). Induction of apoptosis in cultured hepatocytes and in regressing liver by transforming growth factor beta 1. *Proc Natl Acad Sci USA.* 89, 5408-12.
67. Oberhammer, F., Bursch, W., Tiefenbacher, R., Froschl, G., Pavelka, M., Purchio, T. and Schulte-Hermann, R. (1993). Apoptosis is induced by transforming growth factor-beta 1 within 5 hours in regressing liver without significant fragmentation of the DNA. *Hepatology.* 18, 1238-46.
68. Duronio, V., Huber, B. E. and Jacobs, S. (1990). Partial down-regulation of protein kinase C reverses the growth inhibitory effect of phorbol esters on HepG2 cells. *J. Cell. Physiol.* 145, 381-389.
69. Woodworth, C. D., Secott, T. and Isom, H. C. (1986). Transformation of rat hepatocytes by transfection with simian virus 40 DNA to yield proliferating differentiated cells. *Cancer Res.* 46, 4018-4026.
70. Woodworth, C. D. and Isom, H. C. (1987). Regulation of albumin gene expression in a series of rat hepatocyte cell lines immortalized by simian virus 40 and maintained in chemically defined medium. *Molec. Cell. Biol.* 7, 3740-3748.
71. Woodworth, C. D., Krieder, J. W., Mengel, L., Miller, T., Meng, Y. and Isom, H. C. (1988). Tumorigenicity of simian virus 40-hepatocyte cell lines: effect of *in vitro* and *in vivo* passage on expression of liver-specific genes and oncogenes. *Molec. Cell. Biol.* 8, 4492-4501.

11

THE ROLE OF FOLATE, CHOLINE, AND METHIONINE IN CARCINOGENESIS INDUCED BY METHYL-DEFICIENT DIETS

Susanne M. Henning and Marian E. Swendseid

UCLA, School of Public Health
Division of Community Health Sciences
71-297 CHS Los Angeles, California 90095

In humans, deficiencies of choline and methionine are not likely to occur except for an occasional methyl deficiency as evidenced by low plasma choline levels and development of fatty livers in some patients given longterm parenteral nutrition (Burt et al. 1980, Tayek et al. 1990 and Buchman et al. 1991). However, folate deficiency is a problem in many population subgroups such as pregnant women, adolescents, elderly people (Sauberlich 1990 and Joosten et al. 1993), and alcoholics (Shaw et al. 1989). The participation of folate in methionine regeneration and other methyl transfers links folate metabolism closely to methyl metabolism (Krumdieck, 1991). In rats a deficiency of choline and methionine and its effect on hepatocellular carcinoma in the absence of carcinogens has been studied extensively. However study of possible involvement of folate deficency in carcinogenesis mediated through alterations in methyl metabolism has been somewhat neglected. Recent reports indicating an increased risk of cancer and precancerous lesions of various types in humans associated with low folate intakes or low red blood cell folate levels increase the relevance of such investigations (Giovannucci, E. et al. 1993, Butterworth, C.E. et al. 1992).

In this review we will summarize present knowledge concerning the role of folate, choline and methionine in carcinogenesis and show that in rats when folate is withheld from a choline and methionine deficient diet the severity of methyl deficiency symptoms is increased and the developoment of hepatic carcinogenesis is enhanced. The various hypotheses that have been proposed regarding a mechanism of carcinogenesis caused by methyl deficiency will also be described.

1. EFFECTS OF DIETS DEFICIENT IN CHOLINE AND METHIONINE AND THOSE DEFICIENT IN CHOLINE, METHIONINE AND FOLATE

Male rats fed a diet deficient in choline and low in methionine develop hepatocellular carcinomas in the absence of carcinogens (Mikol et al. 1983 and Ghoshal et al.

Dietary Fats, Lipids, Hormones, and Tumorigenesis, edited by Heber and Kritchevsky
Plenum Press, New York, 1996

1984). In these rats livers undergo fat accumulation, cell necrosis, cell death, increased proliferation, cirrhosis and development of GGT-positive foci and finally hepatocellular carcinoma (Ghoshal et al. 1993). While there is a strong indication that methionine/choline deficient diets can serve as effective promoting agents in multistage hepatocarcinogenesis (Giambaresi, 1980), it is controversial as to whether such diets may also cause initiation of individual hepatocytes (Sawada et al. 1990). Ghoshal et al (1987) interpreted the occurence of gamma-GT-positive foci after feeding the Lombardi diet as evidence for initiation because the livers of the control rats fed the choline supplemented diet did not show any foci even after promotion with 2-acetylaminofluorene or partial hepatectomy. Sawada et al (1990), however, assumed that initiation by the diet would be accompanied by a continuing increase in the total number of foci per liver throughout the experimental period. Since they did not show an increase in number of foci over time they proposed that the carcinogenic effect of feeding a methyl deficient diet results more from its strongly promoting effect than from any intitating activity by the diet (Sawada et al 1990). A study of Chandar and Lombardi (1988) suggests an additional late event beyond promotion, because rats fed sequentially the Lombardi diet and a choline-supplemented Lombardi diet developed hepatocellular tumors more frequently (73%) compared to rats fed the Lombardi diet alone (26%).

The diet most frequently used for longterm studies with the methyl deficient rat model is the Lombardi diet (4), which is based on 9% peanut meal, 8% soy protein isolate and 1% casein. It is therefore low in methionine and it is also choline deficient but adequate amounts of folate and vitamin B12 are present. Feeding this diet to young male rats has been shown to cause hepatocellular carcinomas in 20-50% of the rats (Chandar and Lombardi, 1988, Ghoshal and Farber, 1984, Nakae et al., 1992). Chandar and Lombardi found that 26% of rats developed carcinomas after 16 months. When Ghoshal and Farber fed the choline deficient (CD) diet to rats for 13-24 months, and showed 51% of rats developed carcinomas. Nakae et al. (1992) also fed the CD diet and 20% of rats developed hepatocellular carcinomas.

Another diet used for this methyl-deficient cancer model is an amino acid defined diet without choline and low in methionine but with adequate amounts of folic acid and vitamin B12. Nakae et al. (6) compared the amino acid defined, choline-deficient methionine-low diet to the Lombardi diet and showed that 100% of the rats fed the amino acid defined diet for 52 weeks developed hepatocellular carcinomas. The amino acid defined diet was very similar in composition to the CD diet. Their interpretation of the different effects on carcinogenesis of the 2 diets was that the diet providing whole proteins require digestion and polypeptides formed during digestion may be stimulate amino acid absorption and thus increase the methionine intake.

All of these diets contain the RDA of folate and vitamin B12. However it is well known that the metabolic pathways of methionine, choline and folate are interdependent (Krumdieck, 1983). For example, when exogenous methionine and choline are limited in the diet, more folate is needed for the remethylation of homocysteine to form methionine (Krumdieck 1991). An increased turnover of the homocysteine pathway, stimulated by decreased availability of S-adenosylmethionine (SAM), the major intracellular methyl donor, increases the requirement for folate methyl groups and leads to a depletion of total folates in the liver (Selhub et al., 1991). Selhub et al showed that in rats feeding a choline-devoid low-methionine diet decreases not only hepatic choline (<50% of control), betaine (30% of control), methionine (80% of control), S-adenosylmethionine (SAM) (60% of control) but also folate (69% of control) (Selhub et al, 1991) and in rats fed a folate deficient diet hepatic choline and phosphocholine concentrations were decreased (Kim et al, 1994).

A decrease in folate availability in turn may lead to a depletion of folate one-carbon groups which are required for the de novo synthesis of purines and pyrimidines. Therefore an imbalance of the nucleotide pool such as shown by James et al. could contribute to the carcinogenic potential of methyl-donor deficiency (James et al, 1992). James et al (1992) showed that dTMP, dTTP, dGTP and dATP levels were decreased in spleens of rats fed diets deficient in choline, folate and low in methionine.

Since folate is so closely related to the one-carbon metabolism and a choline/methinonine deficiency alone is able to generate a folate deficiency, it seems likely that a combined folate, choline and methionine deficient diet will have a stronger carcinogenic effect compared to either a folate deficient or a preformed methyl-deficient diet. There are a few studies supporting this view. Feeding a folate, choline-deficient, methionine low diet resulted in increased fat accumulation and decreased SAM:SAH ratio in the liver as compared to a diet deficient in only choline and methionine (Henning et al 1989). Tuma et al (1975) also observed an increase in hepatic triglyceride accumulation when the folate antagonist Methotrexate was administered in addition to a choline, methionine-deficient diet with adequate vitamin B12.

Female rats until recently were believed to be resistant to a choline deficient diet (Saito et al (1991). Feeding a choline and folate deficient diet showed no difference in the basic spectrum and sequence of the morphologic alterations to hepatocarcinogenesis in female rats compared to feeding a choline deficient diet (Saito et al. 1994). There was a difference however in the degree and time of onset of the alterations. The more severe the methyl/folate deficiency, the greater the degree and the shorter the time of onset.

There is also growing evidence that folate deficiency alone in rats can act as a cocarcinogen. For example, the incidence of mammary tumors in rats by procarbazine treatment was shown to be significantly increased by methotrexate pretreatment (Rogers et al. 1990). Methotrexate, a inhibitor of dehydrofolate reducatase (EC 1.5.1.4.) results in impaired production of tetrahydrofolate and 1 carbon unit transfer (Barak et al. 1982). Also rats treated with dimethylhydrazine showed a higher incidence of colonic neoplasia when fed a folate-deficient diet (Cravo et al. 1992).

Further, a localized folate deficiency has been implied in the etiology of human epithelial cell cancer in several studies (Krumdieck 1991). For example in a small clinical trial diagnosed cervical dysplasia in oral contraceptive users was shown to improve with folate supplements (Butterworth et al, 1982). In a larger follow up study, however, no improvement of the dysplasia status and no alterations of the course of established disease was observed by an oral foalte supplement of 10 mg/day (Butterworth et al, 1992). In the same study it was shown that the presence of human papilloma virus 16 was less prevalent in women with the highest levels of red blood cell folate (Butterworth, 1992). They concluded that folate deficiency facilitates the incorporation of the viral genome into the host DNA at fragile DNA sites.

In a small trial of heavy smokers with bronchial squamous metaplasia, preliminary evidence was found of a reduction in metaplasia with supplements of folate and vitamin B12 (Heimburger, 1988). Another recent study (Saito et al, 1994) showed that bronchial squamous metaplasia was reduced by administration of folate and vitamin B12.

An epidemiological study found significantly lower levels of vitamin B12 and folate in the blood of patiens with esophageal cellular dysplasia or malignancy compared to cytologically normal controls (Jaskiewicz, 1988).

There is accumulating evidence that folate deficiency is involved in colon cancer. An epidemiological study by Giovannuci et al (1993) assessed the dieatary intake of folate and methionine in women of the Nurses' Health Study and men of the Health Professionals Follow-up Study. They showed that high dietary folate or methionine was inversely associated with risk of colorectal adenoma in women and men and suggested that a methyl group deficient diet may be linked to early stages of colorectal neoplasia. The consumption of alcohol was also associated with the occurrence of distal colon adenoma for men and women. Alcohol itself lowers folate blood levels by affecting folate absorption from the intestine and increase folate cleavage (Shaw et al. 1989). The combination of low folate, low methionine and high alcohol intake resulted in a strong risk factor of 3.2. (Giovannuci et al. 1993). Freudenheim et al (1991) found a decrease in risk associated with kilocalorie-adjusted folate intake for rectal cancer for both males and females. The association for colon cancer was not as consistent nor as strong. Analysis of data from 332 cases of adenomatous polyps and 350 controls showed a protective effect of red blood cell folate concentration against the development of colorectal polyps in men but not in women (Bird et al. 1995).

In humans, hypomethylation has been observed in DNA from colon tissue, both benign colon polyps and malignant carcinoma, compared to adjacent normal tissue (Feinber and Vogelstein 1983 and Goelz et al 1985 and Feinberg et al, 1988). A comparison of normal and neoplastic tissue from the same patient showed hypomethylation in a specific CCGG site in the third exon of c-myc (Sharrard et al 1992).

In our laboratory we developed a folate/methyl-deficient diet, which also led to the typical symptoms of a choline deficient diet, such as fatty livers, decreased hepatic SAM concentration and increased or not altered hepatic SAH levels.

The diet was based on 6% casein 6% gelatin and 15% soyoil. It contained 0.23% methionine, 10 ug vitamin B12/kg diet and no added choline or folate (Table 1). 100% of rats developed hepatocellular carcinomas (Table 2 and Figure 1) in a 12 month period. The diet therefore seems to be more severely methyl deficient as compared to the Lombardi diet. Evidence for oxidative damage was demonstrated by decreased concentrations of plasma and tissue alpha and gamma tocopherol levels (Table 3).

Table 1. Diet composition[1]

	MCFD g/kg	Control[2]
Casein[2]	60	60
Gelatin[3]	60	60
Sucrose[2]	477	477
Alphacell[2]	50	50
AIN76 vitamin mix[2]	10*	10**
AIN 76 mineral[2]	35	35
Soy oil[6]	150	150
L-cysteine[2]	2	2
L-threonine[3]	1.5	1.5
L-tryptophane[4]	0.27	0.27
Corn starch[2]	150	150

* AIN76 vitamin mix without choline and folic acid,

** AIN76 vitamin mix

[1] The methionine and cysteine content of the diet was calculated to be 2.3 g of methionine per kg and 2.3 g of cysteine.

[2] ICN Biochemicals (CLevland, OH). [3] SigmaChemical (St. Louis, MO). [4] Nutritional Biochemicals (Clevland, OH). [5] BioServ (Frenchtown, NJ). [6] Rykoff-Sexton Inc. (Los Angeles, CA).

Table 2. Number and size of tumors developed in rats fed the indicated diet

			# of rats with tumors of indicated size			
Diet # of rats	Per group	None	0-0.5	0.5-0.8	0.8-1.2	1.2-2.0
Control**	3	3				
MCFD	3	1	1(1)* 15 mo		1(2)	
Control	2	2				
MCFD	10		2(10)		5(1),1(2),1(3)	1(1)

*Number indicates the number of rats which developed tumors of the indicated size. Number in parenthesis indicates the number of tumors per rat.
**Control contained 60 g casein and 60 g gelatin, 150 g soy oil, 0.003 g folic acid and 2 g cysteine/kg diet. MCFD was the same as control without folic acid.

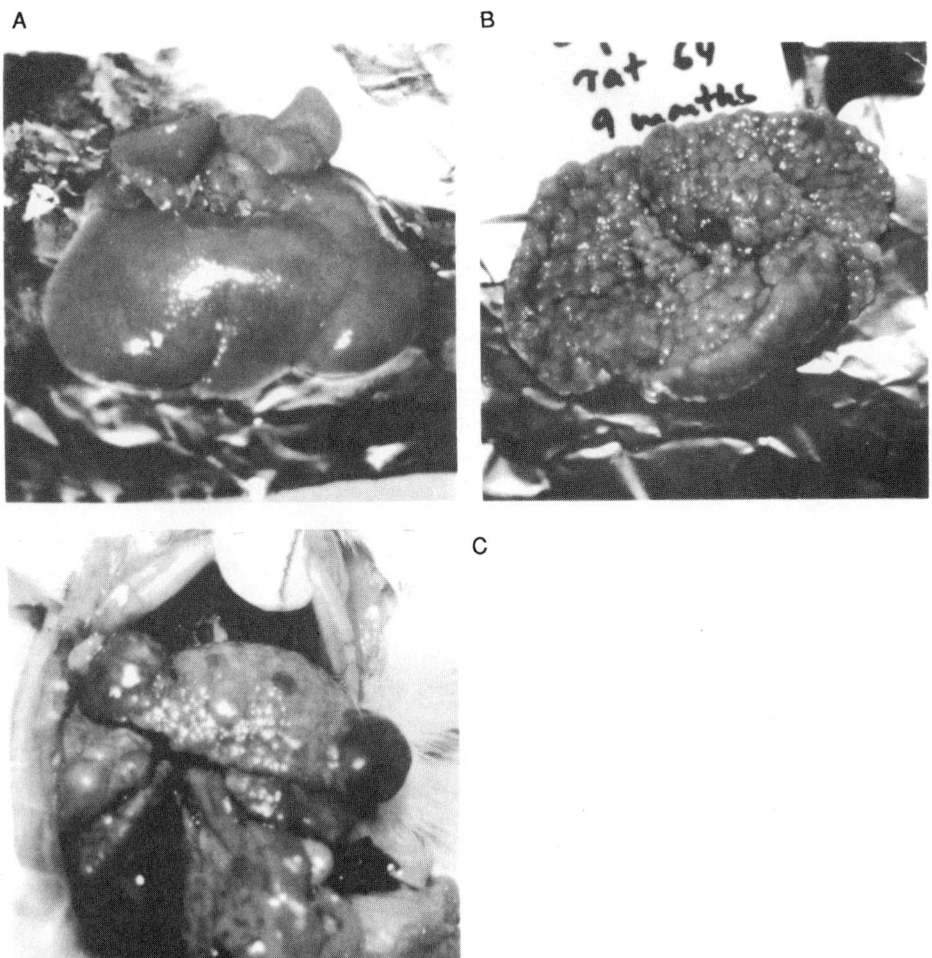

Figure 1. Liver of rats fed A. control diet for 12 months, B. MCFD diet for 9 months and C. MCFD diet for 15 months.

Table 3. Gamma-and Alpha-tocopherol concentration of different tissues and plasma in rats fed the indicated diet

	Tissues (nmol/g tissue)					Plasma
Diet	Liver	Lung	Heart	Spleen	Adipose	(μmol/L)
Gamma-tocopherol						
Control	5.6±2.0[a]	4.9±0.13[a]	5.4±3.0[a]	8.9±1.2[a]	1.3±0.1[a]	1.3±0.3[a]
MCFD	10.8±0.8[a,c]	2.6±0.98[b]	2.1±0.2[b]	7.2±1.6[a]	2.1±0.9[a]	0.5±0.07[b]
Alpha-tocopherol						
Control	182.4±56[a]	70.4±10.7[a]	75.5±6.5[a]	103.0±0.2[a]	11.0±3.2[a,b]	22.5±3.9[a]
MCFD	180.8±22[a]	36.1±12.2[b]	35.0±12.9[b]	69.4±6.4[b]	13.6±7[a]	6.5±2.6[b]

2. HYPOTHESES PROPOSED TO EXPLAIN THE MECHANISM OF CARCINOGENESIS PRODUCED BY METHYL/FOLATE DEFICIENT DIETS

Several hypotheses have been proposed to explain the development of carcinogenesis in rats fed a methyl deficient diet and they will be discussed as follows:

2.1. Oxidative Damage

One of the possible mechanisms for the initiation of hepatocellular carcinoma is the generation of free radicals, which trigger a peroxidation of liver nuclear membrane lipids, and lead to DNA damage. There are many studies that show evidence of oxidative damage and the prevention of damage by antioxidants (Banni et al. 1989, Ghoshal et al. 1990, Nakae et al. 1990, Rushmore et al. 1987). It is controversial if all free radicals originate in the body or if they are also absorbed as part of the diet (Banni et al. 1990). The diet mostly used for study of methyl deficiency induced carcinogenesis is the Lombardi diet with 10% partially hydrogenated fat (Banni et al. 1990) which contains stable fatty acid isomers with conjugated dienes (Banni et al 1989). The use of electron spin resonance spectroscopy however was not successful in demonstrating any abnormal pattern of free radicals in the liver of rats fed a methyl-deficient diet.

Rushmore et al (1984) described the occurence of conjugated dienes in rat liver nuclei as early as 1 day after feeding the methyl deficient diet. The concentration of conjugated dienes peaked at 3 days. At 7 days the level of conjugated dienes was almost back to the inital concentration (Rushmore et al, 1984). Lipid peroxidation (conjugated dienes) in the mitochondria were first dectected at 3 days of feeding the methyl-deficient diet (Rushmore et al, 1987). It increased to a maximum at 28 days and dissappeared after 48 days. Rushmore et al (1987) were also the first to show that the administration of an antioxidant (N-tert-butyl-alpha-phenylnitrone) (PBN) via gastric intubation was able to prevent nuclear lipid peroxidation. An intragastic application of the antioxidant could possibly prevent further oxidation of the dietary fat and reduce the amount of conjugated dienes absorbed based on the evidence presented by Banni et al (1989). Several other studies showed the preventive effect of different antioxidants on the oxidative effects of feeding a methyl-deficient diet (Ghsoshal et al. 1990, Nakae et al 1994, Mizumoto et al. 1994,Denda et al 1994). Ghoshal et al (1990) showed that N-p-methoxyphenylacetyl-dehydroalanine (AD5) was very effective in preventing nuclear lipid peroxidation, but not N.N-diphenyl-p-phenylenediamine (DPPD), butylated hydroxyanisole (BHA) and Trolox C a water-soluble analogue of vitamin E. Nakae et al (1994) demonstrated that mixing DPPD or butylated hydroxytoluene (BHT) in a methyl-deficient diet reduces the size of GSTP-positive lesions without affecting the total number. At the same time TBARS generation was reduced without any effect on the formation of 8-OHdG (Nakae et al 1994). A comparison of the antioxidant effects of ascorbic acid, its lipophilic derivative 2-O-octadecylascrobic acid, alpha-tocopherol and its hydrophilic derivative 6-hydroxy-2,5,7,8-tetramethylchroman-2-carboxylic acid (Trolox) showed that CV-3611 exerted the most pronounced inhibitory effect on the number and size of GGT-positive foci, formation of 8-OHdG and TBARS caused by feeding a methyl-deficient diet (Mizumoto et al 1994).

A more direct indicator of oxidative damage to DNA is the formation of 8-hydroxydeoxyguanosine (8-OHdG) (Nakae et al 1990). Nakae et al (1990) were able to show an association of the development of GGT-positive preneopastic lesions and the concentration of 8-OHdG in rats fed 2 different methyl-deficient diets. An increase in 8-OHdG levels was related to higher number and larger size of GGT-positive foci.

Another approach in demonstrating the role of free radicals in the initiation of hepatocellular carcinoma by feeding a methyl-deficient diet is the evidence that feeding an iron deficient methyl-deficient diet reduced the number of GGT-positive foci and the formation of TBARS and 8-OHdG by about 50-70% (Yoshiji et al 1992). Since *in vivo* iron is know to act as a catalyst in the formation of free radicals via the Fenton reaction (Halliwell and Gutteridge 1984) feeding an iron deficient methyl-deficient diet may lead to a reduction in free radical formation. Yoshiji et al (1992) also point out that this inhibition of lipid peroxidation by iron-deficiency is evidence that lipid peroxidation originates in the body and not in the diet.

In our studies a decrease in plasma vitamin C and tissue alpha and gamma-tocopherol concentration was observed in rats fed a choline, folate-deficient, methionine-low diet (Table 2) (manuscript in preparation). Since these vitamins are important antioxidants, a decrease in their concentrations in plasma and tissues is most likely associated with an increased oxidative stress.

2.2. Hypomethylation and Proto-Oncogene Expression

This hypothesis has been well reviewed by Poirier (1994) and therefore is only briefly described here. This hypothesis proposes that hypomethylation of DNA caused by methyl-deficiency followed by alterations of proto-oncogene or tumor supressor gene expression is a causative factor in hepatocarcinogenesis.

It has been observed that the administration of inhibitors of DNA methylation such as 5-azacytidine, 5-azadeozycytidine or adenosine dialdehyde after carcinogen administration potentiated the initiation of the carcinogenic process (Denda et al 1985 and Antony et al, 1987).

General hypomethylation of liver DNA (Locker et al 1986, Wilson et al 1984, Wainfan et al 1989) and specific hypomethylation of proto-oncogenes such as c-myc, c-fos, c-Ha-Ras and c-K-Ras (Bhave et al 1988, Christman et al 1993) have been documented by several investigators. The hypomethylation was observed as early as several days after feeding a severely methyl/folate-deficient diet (Wainfan et al. 1989) and as late as 14 months in livers and tumors of rats fed the Lombardi diet (Locker et al. 1986). The hypomethylation of DNA coincided with changes in mRNA levels of c-myc, c-fos and c-Ha-ras, but not c-Ki-ras (Dizik et al. 1991). Feeding the Lombardi diet and the same diet supplemented with choline sequentially gave larger gene amplifications of c-myc (Chandar et al. 1989). C-myc amplification was even detected in tumors that arose in rats fed the Lombardi diet for only 3 months. Feeding the Lombardi diet also led to the development of mutant p53 protein in 22 of 27 tumors with immunostaining and 18 of 20 with immunoblotting (Smith et al. 1993). C-DNA sequencing showed that 7 of 11 hepatocellular carcinomas contained point mutations within codon 1120-290 of the gene and 1 microdeletion in the same region. No mutational hot spot was revealed. Mutant p53 was also detected in nontumorous liver tissue of tumor bearing livers and in livers without tumors from animals fed the Lombardi diet for 10 months or longer (Smith et al. 1993).

A first indication of heritable phenotypic changes in hepatocyte leading to tumorigenesis in rats fed a severely methyl-deficient diet was shown by Christman et al (1993). They tested the reversibility of the gene hypomethylation and mRNA levels of proto-oncogenes by feeding a choline-supplemented diet after feeding a severly methyl/folate-deficient diet for 4 weeks (Christman et al. 1993). The overall DNA hypomethylation and gene expression of c-fos and c-myc returned to levels of age matched controls fed the choline supplemented diet, but changes in the methylation pattern persisted in c-fos, c-myc and c-Ha-ras.

The fact that the changes in methylation, gene expression and P53 mutation are not only observed in the tumor tissue but also in the surounding nontumorous liver tissue of tumor bearing livers suggests that this is not the causative link to tumorigenesis and that other events must occur to result in hepatocellular carcinoma induction.

A possible association between oxidative stress and genome expression was shown by several investigators (Abate et al. 1990 and Amstad et al. 1990 + 1992). Abate et al. (1990) found that modification of the redox state of the proto-oncogenes c-fos and c-jun may contribute to the regulation of its DNA binding activity. Amstad et al. (1992) demonstrated that an extracellular burst of active oxygen induced c-fos in mouse epidermal cells JB6.

There is also indication of a genetic difference in the response to the choline/methionine deficient diet (Hinrichsen et al 1991). Male and female PVG/R8 rats fed the Lombardi diet exhibit only minimal fatty infiltration of the liver, no accumulation of hepatic 8-OHdG (Hinrichsen et al. 1991) and minimal formation of GGT-positive foci compared to Fischer 344 rats (Hinrichsen et al 1993). This interstrain variation was not observed when PVG rats were subjected to a standard diethylnitrosamine/phenobarbital treatment for 13-14.5 months (Weinber et al 1987). Possibly the genetic difference might be due to a difference in antioxidant defense or DNA repair.

2.3. Alterations in 1,2-sn-Diradylglycerol and Protein Kinase C Activity

This hypothesis is described in detail by Zeisel et al in this book and by DaCosta et al. (1993). Briefly, it is hypothesized that the development of fatty livers might be associated with the accumulation of 1,2-sn-diraylglycerol and subsequent activation of protein kinase C (DaCosta et al. 1993). Larson and Cerutti (1989) showed that the protein kinase C activity was also enhanced by exposure to oxidants. Since oxidative damage has been frequently observed in choline/methionine deficient rats, oxidative stress might also contribute to the enhancement of protein kinase C as measured by Zeisel et al.

2.4. DNA Damage, Poly (ADP-Ribose)Polymerase Activity and NAD Levels

Based on the observation in our experiments with methyl/folate deficiency mentioned above the following hypothesis explaining the mechanism of carcinogenesis caused by folate/methyl deficient diets was developed. As described earlier, DNA damage can be caused either by imbalance of the nucleotide pool because of a shift of scarce methyl groups to SAM synthesis and away from purine and pyrimidine formation (James et al. 1992), by oxidative damage (Muehlematter et al. 1988), or hypomethylation (Poirier 1994). James et al. (1989) showed evidence of DNA damage by measuring increased amounts of DNA-strand breaks in spleen cells of rats fed a methyl/folate-deficient diet. Rushmore et al. (1986) also showed alkali-labile lesions in livers of rats fed a choline-deficient methionine-low diet.

Table 4. Total liver lipid and NAD concentrations in tissues of Fischer rats fed the indicated diets for 2, 6 and 15 months

Diet*	Liver lipid mg/g	Whole blood μM/L	Liver NAD μmol/g fat free liver	Muscle NAD nmol/g
2 mo				
Control	56±3.8[a]	65±4.4[a]	1114±96[a]	729±27[a]
MCFD	272±10[b]	52±2.8[b]	620±67[b]	769±37[a]
6 mo				
Control	56±14[a]	63±6.3[a]	1195±35[a]	711±65[a]
MCFD	174±16[b]	49±4.1[b]	691±19[b]	723±5[a]
15 mo				
Control	82±14	91±11	1000±288	561±34
MCFD	150±16	53±10	656±98	353±60

* Control diet contained 6% caseine, 6% gelatin, 15% soyoil, 0.42% choline and 0.2% cystine. MCFD diet was same as control without folate.

DNA strand breaks in turn act as stimulating event for poly (ADP-ribose)polymerase (PARP), a nuclear enzyme, which catalyzes the formation of poly ADP-ribose polymers from NAD (Lautier et al. 1993). Its activity is associated with DNA repair and cell regeneration. Muehlematter et al (1988) demonstrated that the PARP activity is increased in cells after exposure to active oxygen. In a previous study we showed that the activty of this enzyme increased in rats fed a mild methyl-deficient diet (Henning, et al. 1991). Rats fed a methyl-deficient diet also showed a decrease in NAD levels in liver, muscle and depending on the severity of the methyl deficiency, also in whole blood (Table 4). Possibly with DNA damage PARP activity is enhanced and NAD levels are decreased. If the NAD level is reduced beyond a certain concentration (about 60% of the control) the PARP activity is decreased (manuscript submitted to Carcinogenesis). We also observed a decrease in hepatic ATP levels in these rats. These observations match the description by Carson et al (1986) of programmed cell death observed in cultured human lymphocytes. He observed DNA damage followed by increased PARP activity, severe NAD and ATP reduction which ended in cell death. We suggest this hypothesis to explain the frequently observed occurrence of cell death in rats fed the choline-deficient, methionine-low diet (Giambarresi et al 1982).

Considering all the information reviewed here we conclude that feeding a folate/choline/methionine deficient diet causes more severe symptoms of a methyl deficiency than a choline/methionine deficient diet. It seems that the folate/choline methionine deficiency acts as a complete carcinogen. The effect of a low folate level on the purin/pyrimidine balance leading to DNA strand breaks and mutations might cause initiation and the increased proliferation and cell death might cause promotion.

REFERENCES

Abate, C., Patel, L., Rauscher III, F.J. & Curran, T. (1990) Redox regulation of Fos and Jun DNA-binding activity *in vitro*. Science 249: 1157-1161.

Amstad, P., Crawford, D., Muehlmatter, D., Zbinden, I., Larsson, R. & Cerutti, P. (1990) Oxidants stress induces the proto-oncogenes, C-fos and C-myc in mouse epidermal cells. Bull Cancer 77: 501-502.

Amstad, P., Krupitza, G. & Cerutti, P.A. (1992) Mechanism of c-fos induction by active oxygen. Cancer Res. 52: 3952-3960.

Antony, A., Rao, P.M., Rajalakshmi, S., & Sarma, D.S.R. (1987) Hypomethylation of DNA during eraly stages of chemical carcinogenesis. Proc. Am. Ass. Cancer Res. 28, 104.

Banni, S., Corongiu, F.P., Dessi, M.A., Iannone, A., Lombardi, B., Tomasi, A., & Vannini, V. Free radicals and lipid peroxidation in liver of rats kept on a diet devoide of choline, Free Rad Res. Comms. 7: 233-240, 1989.

Banni, S., Salgo, M.G., Evans, R.W., Corongiu, F.P. & Lombardi, B. Conjugated diene and trans fatty acids in tissue lipids of rats fed an hepatocarcinogenic choline-devoid diet. Carcinogenesis 11: 2053-2057, 1990.

Barak, A.J., Tuma, D.J. & Beckenhauer, H.C. Methotrexate hepatotoxicity. J. Am. Coll. Nutr. 3: 93-96, 1984.

Bhave, M.R., Wilson, M.J. & Poirier, L.A. (1988) c-H-ras and c-K-ras gene hypomethylation in the livers and hepatomas of rats fed methyl-deficient, amino acid-defined diets. Carcinogenesis 9: 343-348.

Bird, C.L., Swendseid, M.E., Witte, J.S., Shikany, J.M., Hunt, I.F., Frankl, H.D., Lee, E.R., Longnecker, M.P. & Haile, R.W. (1995) Red cell and plasma folate, folate consumption, and the risk of colorectal polyps. Cancer Epidem. Biomarker Prevention (submitted).

Buchman, A., Jenden, D., Roch, M. & Rice, K, (1991) Plasma choline levels in long term TPN patients. FASEB J. 9: A1452.

Burt, M.E., Hanin, L. and Brennan, M.F. (1980) Choline deficiency associated with total parenteral nutrition. Lancet 2, 638-639.

Butterworth, C.E., Hatch, K.D., Macaluso, M., Cole, P., Sauberlich, H.E., Soong, S.-J., Borst, M., Baker, V.V. (1992) Folate deficiency and cervical dysplasia. JAMA 267: 528-533.

Butterworth, C.E., Jr., Hatch, K.D., Mueller, G.H., Krumdieck, C.L. (1982) Improvement in cervical dysplasia associated with folic acid therapy in users of oral contraceptives. Am. J. Clin Nutr. 35: 73-82.

Carson, D.A., Seto, S., Wasson, D.B. and Carrera, C.J. (1986) DNA strand breaks, NAD metabolism, and programmed cell death. Exp. Cell Res., 164, 273-281.

Chandar, N. and Lombardi, B. (1988) Liver cell proliferation and incidence of hepatocellular carcinomas in rats fed consecutively a choline-devoid and a choline-supplemented diet. Carcinogenesis, 9, 259-263.

Chandar, N., Lombardi, B. & Locker, J. (1989) c-myc gene amplification during hepatocarinogenesis by a choline-devoid diet. Proc. Natl. Acad. Sci, USA 86: 2703-2707.

Christman, J.K., Sheikhnejad, G., Dizik, M., Abileah, S. & Wainfan, E. (1993) Reversebility of changes in nucleic acid methylation and gene expression induced in rat liver by severe dietary methyl deficiency. Carcinogenesis 14: 551-557.

Cravo, M.L., Mason J.B., Dayal, Y., Hutchinson, M., Smith, D., Selhub, J. and Rosenberg, I.H. (1992) Folate deficiency enhances the development of colonic neoplasia in dimethylhydrazine-treated rats. Cancer Res. 52: 5002-5006.

DaCosta, K.-A., Cochary, E.F., Blusztajn, J.K., Garner, S.C. and Zeisel S.H. (1993) Accumulation of 1,2-sn-diradylglycerol with increased membrane-associated protein kinase C may be the mechanism for spontaneous hepatocarcinogenesis in choline-deficient rats. J. Biol. Chem. 268: 2100-2106.

Denda, A., Rao, P.M., Rajalakshmi, S., & Sarma, D.S.R. (1985) 5-Azacytidine potentiates initiation induced by carcinogens in rat liver. Carcinogenesis 6: 145-146.

Denda, A., Tang, Q., Endoh, T., Tsujiuchi, T., Horiguchi, K., Noguchi, O., Mizumotot, Y., Nakae, D & Konishi, Y. (1994) Prevention by acetylsalicylic acid of liver cirrhosis and carcinogenesis as well as generation of 8-hydroxydeoxyguanosine and thiobarbituric acid-reactive substances caused by a choline-deficient, l-amino acid-defiened diet in rats. Carcinogenesis 15: 1279-1283.

Dizik, M., Christman, J.K. & Wainfan, E. (1991) Alterations in expression and methylation of specific genes in livers of rats fed a cancer promoting methyl-deficient diet. Carcinogenesis 12: 1307-1312.

Feinberg, A.P. & Vogelstein, B. (1983) Hypomethylation distinguishes genes of some human cancers from their normal counterparts. Nature 301: 89-92.

Feinberg, A.P., Gehrke, C.W., Kuo, K.C. & Ehrlich, M. (1988) Reduced genomic 5-methylcytosine content in human colonic neoplasia. Cancer Res. 48: 1159-1161.

Freudenheim, J.L., Graham, S., Marshall, J.R., Haughey, B.P., Cholewinski, S., Wilinson, G. (1991) Folate intake and carcinogenesis of the colon and rectum. Int. J. Epidemiol. 20: 368-374.

Ghoshal, A.K., Rushmore, T.H., Buc-Calderon, P., Roberfroid, M. & Farber, E. Prevention by free radical scavenger AD5 of prooxidant effects of choline deficiency. Free Radical Biol Med 8:3-7, 1990.

Ghoshal, A.K. and Farber, E. (1984) The induction of liver cancer by dietary deficiency of choline and methionine without added carcinogens. Carcinogenesis, 5, 1367-1370.

Ghoshal, A.K., Rushmore, T.H. & Farber, E. (1987) Initiation of carcinogenesis by a dietary deficiency of choline in the absence of added carcinogens. Cancer Let. 36: 289-296.

Ghoshal, A.K. and Farber, E. (1993) Biology of disease. Choline deficiency, lipotrope deficiency and the develoment of liver disease including liver cancer: a new perspective. Lab. Inv., 68, 255-260.

Ghoshal, A.K., Rushmore, T.H., Buc-Calderon, P., Roberfroid, M. and Farber, E. (1990) Prevention by free radical scavenger AD5 of peroxidant effects of choline deficiency. Free Rad. Biol. Med. 8, 3-7.

Giambarresi, L.K., Katyal, S.L. and Lombardi, B. (1982) Promotion of liver carcinogenesis in the rat by a choline-devoid diet: role of liver cell necrosis and regeneration. Br. J. Cancer, 46: 825-829.

Giovannucci, E., Stampfer, M.J., Colditz, G.A., Rimm, E.B., Trichopoulos, D., Rosner, B.A., Speizer, F.E., Willett, W.C. (1993) Folate, methionine, and alcohol intake and risk of colorectal adenoma. J. Nat. Cancer Inst. 85:875-884.

Goelz, S.E., Vogelstein, B., Hamilton, S.R. & Feinberg, A.P. (1985) Hypomehtylation of DNA from bengin and malignant human colon neoplasms. Science 228: 187-190.

Halliwell, B. & Gutteridge, J.M.C. (1984) Oxygen toxicity, oxygen radicals, transition metals and disease. biochem. J. 219: 1-14.

Heimburger, D.C., Alexander, C.B., Birch, R., Butterworth, C.E., Bailey, W.C., Krumdieck, C.L. (1988) Improvement in bronchial squamous metaplasia in smokers treated with folate and vitamin B12. JAMA 259:1525-1530.

Henning, S.M., McKee, R.W. and Swendseid, M.E. (1989) Hepatic poly (ADP ribose) polymerase activity in methyl donor- deficient rats. J. Nutr., 119, 1528-1531.

Henning, S.M., McKee, R.W. and Swendseid, M.E. (1989) Hepatic content of S-adenosylmethionine, S-adenosylhomocysteine and glutathione in rats receiving treatments modulationg methyl donor availability. J. Nutr., 119, 1478-1482.

Hinrichsen, L., West, M., Floyd, R. & Sudilovsky, O. (1991) Liver 8-hydroxydeoxyguanosine (8-OHdG) is not increased in PVG rats, resistant to promotion by a choline-devoid (CD) diet. Proc. Am. Assoc. Cancer Res. 32: 899A.

Hinrichsen, L.I., Sudilovsky, E.C., Roggero, E., Jiang, S.-H. & Sudilovsky, O. (1993) Differntial responses of two inbred rat strains to a choline-deficient diet during liver carcinogenesis. Carcinogenesis 14: 1951-1956.

James, S.J., Cross, D.R. and Miller B.J. (1992) Alterations in nucleotide pools in rats fed diets deficient in choline, methionine and/or folic acid. Carcinogenesis, 13, 2471-2474.

James, S.J., Yin, L. and Swendseid, M.E. (1989) DNA strand break accumulation, thymidylate synthesis and NAD levels in lymphocytes from methyl donor-deficient rats. J. Nutr., 119, 661-664.

Jaskiewicz, K., Marasas, W.F.O., Lazarus,C., Beyers, A.D., Van Helden, P. D. (1988) Association of esophageal cytological abnormalities with vitamin and lipotrope deficiencies in populations at risk for esophageal cancer. Anticancer Res. 8: 711-716.

Joosten, E., Van den Berg, A., Riezler, R., Naurath, H.J., Lindenbaum, J., Stahler, S.P. & Allen, R.H. (1993) Metabolic evidence that deficiencies of vitamin B-12 (cobalamin), foalte, and vitamin B-6 occur commonly in elderly people. Am. J. Clin. Nutr. 58: 468-476.

Kim Y-I., Miller, J.W., DaCosta, K-A., Nadeau, M., Smith, D., Selhub, J., Zeisel, S.H. and Mason, J.B. (1994) Severe folate deficiency causes secondary depletion of choline and phosphocholine in rat liver. J. Nutr. 124, 2197-2203, 1994.

Krumdieck, C.L. (1991) Localized folate deficiency and cancer. In: Vitamins and Cancer prevention (eds.), Wiley-Liss Inc. pp39-49.

Larson, R. & Cerutti, P. (1989) Translocation and enhancement of phosphotransferase activity of protein kinase C following exposure of mouse epidermal cells to oxidants. Cancer Res. 49: 5627-5632.

Lashner, B.A. (1993) kRed blood cell folate is associated with the development of dysplasia and cancer in ulcerative colitis. J. Cancer Res. Clin. Oncol. 119: 549-554.

Lautier, D., Lagneux, J., Thibodeau, J., Menard, L. and Poirier, G.G. (1993) Molecular and biochemical features of poly (ADP-ribose) metabolism. Mol. Cell. Biochem., 122, 171- 193.

Locker, J., Reddy, T.V. & Lombardi, B. (1986) DNA methylation and hepatocarcinogenesis in rats fed a choline-devoid diet. Carcinogenesis, 7: 1309-1312.

Mikol,Y.B., Hoover, K.L., Creasia, D. and Poirier, L.A.(1983) Hepatocarcinogenesis in rats fed methyl-deficient L-amino defined diet in rats. Cancer Res., 52, 5042-5045.

Mizumoto, Y., Nakae, D., Yoshiji, H., Andoh, N., Horiguchi, K., Endoh, T., Kobayashi, E., Tsujiuchi, T., Shimoji, N., Denda, A., Tsujii, T., Nagao, M., Wakabayashi, K. & Konishi, Y. Inhibitory effects of 2-O-octadecylascorbic acid and other vitamin C and E derivatives on the induction of enzyme-altered putative preneoplastic lesions in the livers of rats fed a choline-deficient, L-amino acid-defined diet. Carcinogenesis 15: 241-246, 1994.

Muehlematter, D., Larsson, R. & Cerutti, P. Active oxygen induced DNA stran breakage and poly ADP-ribosylation in promotable and non-promotable JB6 mouse epidermal cells. Carcinogenesis 9: 239-245, 1988.

Nakae, D., Yoshiji, H., Mizumoto, Y., Horiguchi, K., Shiraiwa, K., Tamura, K., Denda, A. and Karishi, Y. (1992) High incidence of hepatocelluar carcinomas induced by a choline deficient L-amino acid defined diet in rats. Cancer Res. 52: 5042-5045.

Nakae, D., Yoshiji, H., Maruyama, H., Kinugasa, T., Denda, A., & Konishi, Y. Production of both 8-hydroxy-deoxyguanosine in liver DNA and gamma-glutamyltransferase-positive hepatocellular lesions in rats given a choline-deficient, L-amino acid-defined diet.

Jpn.J.Cancer Res. 81:1081-1084,1990.

Nakae, D., Mizumoto, Y., Yoshiji, H., Andoh, N., Horiguchi, K., Shiraiwa, K., Kobayashi, E., Endoh, T., Shimoji, N., Tamura, K., Tsujiuchi, T., Dend, A. & Konishi, Y. Different roles of 8-hydroxydeoxy-guanosine formation and 2-thiobarbituric acid-reacting substance generation in the early phase of liver carcinogenesis induced by a choline-deficient, l-amino acid-defined diet in rats. JpnJ.Cancer Res. 85: 499-505, 1994.

Poirier, L.A., (1994) Methyl group deficiency in hepatocarcinogenesis. Drug Metabol. Reviews 26: 185-199.

Rogers, A.E., Akhtar, R., Zeisel, S.H. (1990) Procarbazine carcinogenicity in methotrexate-treated or lipot-rope-deficient male rats. Carcinogenesis 11: 1491-1495.

Rushmore, T.H., Lim, Y.P., Farber, E. & Ghoshal, A.K. Rapid lipid peroxidation in the nuclear fraction of rat liver induced by a diet deficient in choline and methionine. Cancer Letters, 24: 251-255, 1984.

Rushmore, T.H., Farber, E., Ghoshal, A.K., Parodi, S., Pala, M. and Taningher, M. (1986) A choline-devoid diet, carcinogenic in the rat, induces DNA damage and repair. Carcinogenesis, 7, 1677-1680.

Rushmore, T.H., Ghazarian, D.N., Subrahmanyan, V., Farber, E. & Ghoshal AmK. Probable free radical effects on rat liver nuclei during early hepatocarcinogenesis with a choline-devoid low methionine diet. Cancer Res 47: 6731-6740, 1987.

Saito, R., Palomba, L., Rao, K.N. & Lombardi, B. (1991) resistance of female Fischer-344 rats to the hepatonecrogenic and hepatocarcinogenic actions of a choline-devoid diet. Carcinogenesis 12: 1451-1457.

Saito, R., Jahnke-Spinnenweber, E., Shinozuka, H. & Lombardi, B. (1994) On the role of compensatory mitogenesis in the hepatocarcinogenicity of choline and multiple-lipotrope devoid diets. Carcinogene-sis 15: 1413-1419.

Saito, M., Kato, H., Tsuchida, T., Konaka, C. (1995) Chemoprevention effects on bronchial squamous metaplasia by foalte and vitamin B12 in heavy smokers. Chest 106: 496-499.

Sauberlich, H.E. (1990) Evaluation of folate nutrition in population groups. In: Folic acid metabolism in health and disease (eds) Wiley-Liss Inc. p. 211-235.

Sawada, N., Poirier, L., Moran, S., Xu, Y.-H. & Pitot, H.C. (1990) The effect of choline and methionine deficiencies on the number and volume of dietylnitrosamine initiatio in rat liver. Carcinogenesis 11: 273-281.

Selhub, J., Seyoum, E., Pomfret, E.A.and Zeisel, S.H. (1991) Effects of choline deficiency and methotrexate treatment upon liver folate content and distribution. Cancer Res., 51, 16- 21.

Sharrard, R.M. Royds, J.A., Rogers, S. & Shorthouse, A.J. (1992) Patterns of methylation of the c-myc gene in human colorectal cancer progression. Br. J. Cancer 656: 667-672.

Shaw, S., Jayatilleke, E., Herbert, V., & Colman, N. (1989) Cleavage of folates during ethanol metabolism. Role of acetaldehyde/xanthine oxidase-generated superoxide. Biochem. J. 257: 277-280.

Smith, M.L., Yeleswarapu, L., Scalamaogna, P., Locker, J. & Lombardi, B. (1993) p53 mutations in hepato-cellular carcinomas induced by a choline-devoid diet in male Fischer 344 rats. Carcinogenesis 14: 503-510.

Tayek, J.A., Bistrian, B., Sheard, N.F., Zeisel, S.H. and Blackbum, G.L. (1990) Abnormal liver function in malnurished patients receiving total parenteral nutrition: a prospective randomized study. J.Am. Coll. Nutr. 9, 76-83.

Tuma, D.J., Barak, A.J., Sorrell, M.F. (1975) Interaction of methotrexate with lipotropic factors in rat liver. Biochem. Pharmacol. 24: 1327-1331.

Wainfan, E., Dizik, M., Stender, M., Christman, J.K. (1989) Rapid appearance of hypomethylated DNA in livers of rats fed cancer-promoting, methyl-deficient diets. Cancer Res. 49: 4094-4097.

Weinberg, W.C., Berkvits, L. & Iannaccone, P.M. (1987) The clonal nature of carcinogen-induced altered foci of y-glutamyl transpeptidase expression in rat liver. Carcinogenesis 8:565-570.

Wilson, M.J., Shivapurkar, N. & Poirier, L.A. (1984) Hypomethylation of hepatic nuclear DNA in rats fed with a carcinogenic methyl-deficient diet. Biochem. J. 218: 987-990.

Yoshiji, H., Nakae, D., Mizumoto, Y., Horiguchi, K., Tamura, K., Denda, A., Tsujii, T. & Konishi, Y. Inhibitory effect of dietary iron deficiency on inductions of putative preneoplastic lesions as well as 8-hydroxy-deoxyguanosine in DNA and lipid peroxidation in the livers of rats casued by exposure to a choline-deficient L-amino acid defined diet. Carcinogenesis 13: 1227-1233, 1992.

Zhang, J., Henning S.M. and Swendseid, M.E. (1993). NAD levels in tissues of methyl-folate deficient rats. FASEB J., A 727.

Zhang, J.Z., Henning, S.M. and Swendseid, M.E. (1993) Poly (ADP-ribose) polymerase activity and DNA strand breaks are affected in tissues of niacin-deficient rats. J. Nutr., 123, 1349-1355.

12

DIETARY FAT AND COLON CANCER

J. R. Hecht

Digestive Disease Center
200 U.C.L.A. Medial Plaza
Suite 365A
Los Angels, California 90095

1. INTRODUCTION

It is estimated that there will be an estimated 149,000 new cases of colorectal carcinoma in the United States this year. The colon is the second most common site of serious malignancy and is the most common in women over 75, the fastest growing segment of the population.[1] Despite intense efforts at early detection, the disease is frequently metastatic at the time of diagnosis. The death rate has hovered at approximately 50% for the past several decades and nearly 60,000 deaths are expected this year. Improved prevention is crucial to reducing the immense human and financial costs of this disease. Study of the link between dietary fat and colorectal carcinogenesis may lead to improved methods of cancer prevention.

2. EPIDEMILOGY

Colorectal carcinoma, like many malignancies, has a significant variation in incidence between different parts of the world. Rates are much higher in the United States and Europe than in Asia, Africa, and South America. Immigrants from low risk countries to high risk areas have higher rates of colorectal carcinoma than their original countrymen.[2] These findings lead to the examination of various dietary and environmental factors which showed a strong correlation between average dietary fat intake and the rate of colorectal cancer at the national level.

This relationship has been corroborated by several other lines of evidence. Members of certain religious groups, such as the Seventh Day Adventists who have a low fat ovolactarian diet, have a decreased incidence of colorectal cancer than do their neighbors.[3] A number of case-control studies have also shown a positive relationship between total fat intake and the risk of developing colorectal cancer in Canada,[4] Australia[5], Utah[6], and New York.[7] There are many possible confounding factors in such studies and it should be noted that these results have not always been confirmed.[8,9]

Dietary Fats, Lipids, Hormones, and Tumorigenesis, edited by Heber and Kritchevsky
Plenum Press, New York, 1996

3. ANIMAL STUDIES

Animal models of colorectal carcinoma have produced even stronger evidence linking fat intake to colorectal carcinogenesis. The most frequently employed model examines the effect of diet on the incidence or number of tumors in rats induced by a carcinogen such as dimethylhydrazine, or its metabolite azoxymethane. Animals fed high fat diets have higher rates of colonocyte proliferation,[10] a marker for carcinogenesis, and develop a greater number of benign and malignant intestinal tumors.[11,12,13,14] As will be discussed later the type of fat as well as the amount ingested is critical to tumor formation.

4. TYPE OF FAT

The type of fat may actually be as important as the quantity of fat in the diet. Epidemiologic studies have attempted to distinguish the effects of different types of dietary fat on colorectal carcinogenesis. Unfortunately, the tools available such as food diaries and recall quetionaires are often not able to separate out the effects of various types of fat from other variables such as total fat, total calories, and other food components. The best data, once again, have come from animal studies.

It is possible to divide dietary fat into three major groups: saturated fats without double bonds, monounsaturated fats with a single double bond, and polyunsaturated fats with more than one double bond. The primary sources of saturated fats are animal fats, monounsaturated fats are found in animal fats and certain vegetable oils such as olive oil, while vegetable oils such as corn oil are good sources of polyunsaturated fats.[15] The relationship between saturated fat and carcinogenesis is not clear. Diets high in saturated fat in the form of lard or polyunsaturated fat in the form of corn oil appear to increase carcinogenesis in animal models.[12,16] A recent large prospective study among 88,751 women in the United States, however, found a significant correlation between the consumption of animal fat and the risk of colon cancer.[17] Interestingly, the risk was more strongly associated with red meat than saturated fat consumption.

The polyunsaturated fat in corn oil and most vegetable oils is in the ω-6 form.[15] Marine fish oils, however, are high in ω-3 polyunsaturated fatty acids. Populations with diets high in marine oils have relatively low rates of cancer despite high fat intakes.[18] While diets high in as ω-6 polyunsaturated fats such as corn or safflower oils increase tumorigenesis in carcinogen treated animals, those high in fish oil do not.[14,19] Other sources of ω-3 polyunsaturated fatty acids such as perilla oil seem to have the same effect.[20] Fish oils decrease inflammation in patients with ulcerative colitis[21] and the rate of colonocyte proliferation in patients with a history of polyps[22] both of which have been associated with the risk of cancer. Marine oils also suppress the formation of colon cancer metastasis in animal models as well.[23]

5. POTENTIAL ETIOLOGIES

Several theories have been proposed to explain these effects of dietary fat, though none have been substantiated. One possible mechanism is by altering the amount and types of bile acids produced. Primary bile acids are synthesized from cholesterol and metabolized to secondary bile acids by anerobic bacteria.[24] Increased serum cholesterol has been associated with an increased risk of colonic adenomas,[25] which are precursors to cancer, and in a series of 92,000 Swedes, with cancer itself.[26] Diets high in fat increase the excretion of bile acids.[27,428,29] Increased fecal bile acid excretion has been found in patients with colonic polyps and cancer.[30,31,32] In animal studies, bile acids have been shown to act as tumor promoters by increasing the turnover of colonic epithelial cells.[33,34,35,36,37,38,39] Secondary bile acids enhance the tumor promoting activities of primary acids.[40] Bacterial flora with increased capacity for secondary bile acid formation have been found in populations with high rates of colon cancer[41] and in omnivores compared to vegetarians.[42,43] Decresing beef fat intake has been associated with a lower level of activity of these bacterial enzymes.[44,45]

Another possible mechanism for the link between fat and colorectal cancer is by alteration of the production of prostaglandins. Prostaglandins are products of the metabolism of the polyunsaturated fatty acid, arachidonic acid, with many different and potent effects on the inflammatory, cardiovascular, hematologic, reproductive, gastrointestinal, renal, and pulmonary systems.[46] The prostaglandin biosynthetic pathway actually begins with with the polyunsaturated fatty acid linoleic acid. While some arachidonate is available in the diet, the majority is synthesized from dietary linoleic acid[47] which is stored in membrane phospholipids.[48] Arachidonic acid then is liberated from the membrane by phospholipases.[49] Colonic mucosal lipid composition in animals and in man can be altered by changes in dietary fatty acid intake.[50,51] We found that increasing dietary linoleic acid increased mucosal linoleic and arachidonic acid content.[52] (Table 1) Substitution of ω-3 fatty acids, such as those found in marine oils, for arachidonic acid alters the formation of prostaglandin-like products which may explain some of the effects of these oils. [53]

Arachidonic acid is metabolized by the enzyme cyclooxygenase (COX, prostaglandin endoperoxide synthase, or PGHS) and the products of this reaction undergo further metabolism into other prostaglandins, thromboxanes, or prostacyclin. The mechanism of action of aspirin and the large group of compounds known as the nonsteroidal

Table 1. Patient BK

	Serum (μM/L Serum)			Tissue (μM/g Colonic Tissue)		
Fatty Acid	Baseline	8 Weeks Diet	% Change	Baseline	8 Weeks Diet	% Change
14:0	305	259	-17.76		79	
14:1w5						
16:0	4289	4292	0.07	1540	1635	5.81
16:1w9	440	428	-2.80	116	85	-36.47
18:0	1326	1465	9.49	889	1074	17.23
18:1w9	3293	3091	-6.54	1572	1502	-4.66
18:2w6	5303	6449	17.77	1332	1664	19.95
20:0	114	57	-100.00			
18:3w9	465	339	-37.17			
20:1w9						
20:2w6						
20:3w9	286	358	20.11		133	
22:0	115	110	-4.55			
20:4w6	1515	1759	13.87	684	875	21.83
22:1w9						
24:0		83			68	
24:1w9	139	136	-2.21		100	
22:6w3	277	244	-13.52		132	
Sum	17867	19070	6.73	6133	7347	19.79
Sum SFA	6149	6266	1.87	2429	2856	14.95
Sum MUFA	3872	3655	-5.94	1688	1687	-.06
Sum PUFA	7846	9149	14.24	2016	2804	28.10

anti-inflammatory drugs (NSAIDs) is inhibition of cyclooxygenase, which leads to decreases in both prostaglandins and inflammation.[54]

Two forms of cyclooxygenase, known as COX-1 and COX-2 have been described.[55] COX-1 appears to be constitutively expressed in almost all cell types and is much less affected by mitogen treatment. COX-2 is stimulated by a variety of mitogens, such as phorbol esters which stimulate protein kinase C, forskolin which acts through protein kinase A, serum, and peptide growth factors that signal through tyrosine kinase.[55]

Prostaglandin E (PGE) and PGF levels in human colonic carcinomas are elevated compared to normal tissue.[56,57] Prostaglandin levels also are elevated in the colonic mucosa of carcinogen treated rats.[58] In addition, the PGE_2 content of premalignant adenomatous polyps is intermediate between that of normal-appearing mucosa and colon cancer tissue.[59] Colorectal cancers have increased levels of the prostaglandin precursor, arachidonic acid, as compared to adjacent mucosa.[60] Intestinal cells in culture overexpressing COX-2 have a more malignant phenotype and inhibition of apoptosis.[61] Human colon cancers have increased expression of COX-2 protein compared to normal tssues.[62] Prostaglandins also appear to be important in regulation of cell growth.

Growth of cell lines derived from colon,[63,64] as well liver[65] and breast[66] carcinomas, is inhibited by aspirin and indomethacin, inhibitors of prostaglandin synthesis, which may induce a reversible arrest of cells in the G_1 phase of the cell cycle.[67] The NSAIDs indomethacin,[68,69] piroxicam,[70,71] and sulindac[72,73] have been shown to decrease the incidence and number or tumors in dimethylhydrazine-induced colorectal cancer in rats. The growth of human colon tumors explants in nude mice also is inhibited by indomethacin.[74]

Several large epidemiologic studies have shown that NSAIDs reduce of the risk of developing colorectal adenomatous polyps and cancers by up to 50%.[75,76,77,78] Perhaps the most striking results have been seen in familial polyposis coli. This dominantly inherited disorder, caused by mutations of the APC gene, is characterized by the formation of up to thousands of adenomatous polyps and invariably leads to colorectal cancer unless the colon is removed.[79,80] Waddell and others showed that in patients with familial polyposis who had remnant colonic mucosa, treatment with sulindac resulted in the disappearance of some adenomatous polyps.[81,82] Recently, phospholipase A_2, critical in releasing arachidonic acid for the prostglandin synthesis pathway, has been shown to modulate the effects of the APC gene in mice.[83]

Unfortunately the use of NSAIDs may result in serious adverse effects on kidney function and gastric mucosal integrity.[84] The importance of COX-2 in these organs was shown by knocking out COX-2 in mice which results in severe renal pathology.[85] The mechanism of the beneficial effects of NSAIDs on colonic neoplasms remains unclear.

6. SUMMARY

Dietary fat in general, and perhaps animal and polyunsaturated fats in particular, appear to increase colon carcinogenesis in animal models and epidemiologic studies. This observation holds the potential to shed light on the underlying mechanisms of colorectal carcinogenesis and reduce morbidity and mortality from the disease by dietary and chemoprevention. While research continues into the relationship between prostglandins and other putative mediators of the effects of fats on the colon, we can suggest that a diet which protects against colorectal cancer would be low in fat with most of that fat coming from vegetable sources. Prospective dietary trials are ongoing which may strengthen or modify these preliminary recommendations.

REFERENCES

1. Wingo PA, Tong T, Bolden S. Cancer statistics, 1995. CA Cancer J Clin. 1995;45:8-30.
2. Wynder EL, Shigematsu T. Environmental factors of cancer of the colon and rectum. Cancer. 20:1520-61, 1967.
3. Phillips RS. Role of life style and dietary habits in risk of cancer among Seventh-Day Adventists. Cancer Res. 35:3513-22, 1975.
4. Jain M, Cook GM, Davis FG, et al. A case-control study of diet and colo-rectal cancer. Int J Cancer. 26:757-68, 1980.
5. Kune S, Kune GA, Watson LF Case-control study of dietary etiological factors: the Melbourne colorectal cancer study. Nutr Cancer. 9:21-42, 1987.
6. Slattery ML, Schumacher MC, Smith KR, West DW, Abd-elghany. Physical activity, diet, and risk of colon cancer in Utah. Am J Epidemiol. 128: 989-99, 1988.
7. Graham S, Marshall J, Haughey B, et al. Dietary epidemiology of cancer in western New York. Am J Epidemiol. 128: 490-503, 1988.
8. Stemmerman GN, Heilbrun LK, Nomura AMY. Association of diet and other factors with adenomatous polyps of the large bowel: a prospective study. Am J Clin Nutr. 47:312-7, 1988.
9. Mc Keigue PM, Adelstein AM, Marmot MG, et al. Diet and fecal steroid profile in a South Asian population with a low cancer rate. Am J Clin Nutr. 50:151-4, 1989.
10. Goettler D, Rao AV, Bird RP. The effects of a low-risk diet on cell proliferation and enzymatic parameters of preneoplastic rat colon. Nutr Cancer. 1987;10:149-62.
11. Nigro ND, Singh DV, Campbell RL, and Pak, RL. Effect of dietary beef fat on intestinal tumor formation by azoxymethane in rats. J Natl Cancer Inst. 1975; 54:429-42.
12. Reddy BS. Dietary fat and colon cancer: animal models. Prevent Med. 1987;16:460-7.
13. Zhao LP. Kushi LH, Klein RD, Prentice RL. Quantitative review of studies of dietary fat and rat colon carcinoma. Nutr Cancer. 1991;15:169-77.
14. Lindner MA. A fish oil diet inhibits colon cancer in mice. Nutr. Cancer. 1991;15:1-11.
15. Dupont J, White P, Carpenter MP, et al. Food uses and effects of corn oil. J Am Coll Nutr. 1990; 9:438-70,.
16. Reddy BS and Maeura Y. Tumor promotion by dietary fat in azoxymethane-induced colon carcinogenesis in female F344 rats: influence of amount and source of dietary fat. JNCI. 72:745-50, 1974.
17. Willett WC, Stampfer MJ, Colditz, GA, Rosner B, Speizer FE. Relationship of meat, fat, and fiber intake to the risk of colon cancer in a prospective study among women. N Engl J Med. 323:1664-72, 1990.
18. Blot WJ, Lanier A, Fraumeni JF Jr. Cancer mortality among Alaskan natives, 1960-1969. J Natl Cancer Inst. 1975; 55:546-54. (CHECK)
19. Reddy BS, Burill C, Rigotty J. Effect of diets high in omega-3 and omega-6 fatty acids on initiation and postinitiation stages of colon carcinogenesis. Cancer Res. 51:487-91, 1991.
20. Narisawa T, Takahashi M, Kotanagi H, et al . Inhibitory effect of dietary perilla oil rich in the n-3 polyunsaturated fatty acid alpha-linoleic acid on colon carcinogenesis in rats. Jpn J Cancer Res. 82:1089-96, 1991.
21. Stenson WF, Cort D, Rodgers J. Dietary supplementation with fish oil in ulcerative colitis. Ann Intern Med. 1992;116:609-14.
22. Anti M, Marra G, Armelao F. Effect of w-3 fatty acids on rectal mucosal cell proliferation in subjects at risk for colon cancer. Gastroenterol. 1992;103:883-91.
23. Cannizzo F, Broitman SA. Postpromotional effects of dietary marine or safflower oils on large bowel or pulmonary implants of CT-26 in mice. Cancer Res. 1989;49:4289-94.
24. Hoffmann AF. The enterohepatic circulation of bile acids in health and disease. In: Sleisenger MH and Fordtran JS, eds. Gastrointestinal disease: pathophysiology, diagnosis, management. Philadelphia: W. B. Saunders, 1989:144-161.
25. Mannes GE, Maier A, Thieme C, Wiebecke B, and Paumgartner G. Relation between the frequency of colorectal adenoma and the serum cholesterol level. N Engl J Med. 1986:315:1634-8.
26. Tornberg SA, Holm L-E, Carstensen JM, Eklund GA. Risks of cancer of the colon and rectum in relation to serum cholesterol and beta-lipoprotein. N Engl J Med. 1986;315:1629-33.
27. Hill MJ, Draser BS, Aries V, et al. Bacteria and aetiology of cancer of the large bowel. Lancet. 1977; 1:95-100.
28. Reddy BS. Dietary fat and its relationship to large bowel cancer. Cancer Res. 1981; 41:3700-5.
29. Hill MJ, DraserBE, Williams REO, et al. Faecal bile-acids and clostridia in patients with cancer of the large bowel. Lancet. 1975;1:535-8.
30. Linos DA, Beard CM, O'Fallon W. Cholecystectomy and carcinoma of the colon. Curr Surg. 1981; 38:391-3.

31. Reddy BS, Mastromarino A, Wynder EL. Further leads on metabolic epidemiology of large bowel cancer. Cance Res. 1975;35:3403-6.
32. Reddy BS and Wynder EL. Metabolic epidemiology of colon cancer: fecal bile acids and neutral sterols in colon cancer patients with adenomatous polyps. Cancer. 1977; 39:2533-9.
33. Chomchai C, Bhadrachari N, Nigro ND. The effect of bile on the induction of experimental intestinal tumors in rats. Dis Colon Rectum. 1974;17:310-2.
34. Reddy BS, Narasawa T, Weisburger JH, Wynder EL. Promoting effect of deoxycholic acid on colonic adenocarcinomas in germfree rats. J Natl Cancer Inst. 1976;56:441-2.
35. Reddy BS and Watanabe K. Effect of cholesterol metabolites and promoting effect of lithocholic acid on colon carcinogenesis in germfree and conventional F344 rats. Cancer Res. 1979;39:1521-4.
36. Narisawa T, Magadia NE, Weisberger JH, Wynder EL. Promoting effect of bile acid on colon carcinogenesis after intrarectal installation of N-methyl-N′-nitro-nitrosoguanidine in rats. J Natl Cancer Inst. 1974;53:1093-7.
37. Cohen BI, Raicht RF, Deschner EE, et al. Effect of cholic acid feeding on N-methyl-N-nitrosourea-induced colon tumors and cell kinetics in rats. J Natl Cancer Inst. 1980;64:573-6.
38. Bagheri SA, Bolt MG, Boyer JL, Palmer RH. Stimulation of thymidine incorporation in mouse liver and biliary tract epithelium by lithocholate and deoxycholate. Gastroenterol. 1978;7:188-92.
39. Ranken, R, Wilson R, Bealmer PM. Increased turnover of intestinal mucosal cells of germfree mice induced by cholic acid. Proc Soc Exp Biol Med. 1971;138:270-2.
40. Reddy BS and Wynder EL. Metabolic epidemiology of colon cancer: fecal bile acids and neutral sterols in colon cancer patients with adenomatous polyps. Cancer. 1977; 39:2533-9.
41. Reddy BS. Dietary fat and its relationship to large bowel cancer. Cancer Res. 1981; 41:3700-5.
42. Goldin BR, Swenson L, Dwyer J, Sexton M, Gorbach SL. Effect of diet and lactobacillus acidophilus supplements on human fecl bacterial enzymes. J Natl Cancer Inst. 1980;64:255-61.
43. Turjman N, Goodman GT, Jaeger B, Nair PP. Diet, nutrition intake, and metabolism in populations at high and low risk for colon cancer: Metabolism of bile acids. Am J Clin Nutr. 1984;40(suppl):937-41.
44. Goldin BR, Swenson L, Dwyer J, Sexton M, Gorbach SL. Effect of diet and lactobacillus acidophilus supplements on human fecl bacterial enzymes. J Natl Cancer Inst. 1980;64:255-61.
45. Reddy BS, Hanson D, MangatB, et al. Effect of high-fat, high-beef diet and of mode of cooking of beef in the diet on fecal bacterial enzymes and fecal bile acids and neutral sterols. J Nutr. 1980;110:1880-7.
46. Moore PK. Prostanoids. New York: Cambridge University Press, 1985. 68-235.
47. Wahle, KW. Dietary regulation of essential fatty acid metabolism and membrane phospholipid composition. Biochem Soc Trans. 18:775-8, 1990.
48. Murphy RC. Biosynthesis and metabolism. In: Watkins WD, et al. Prostaglandins in Clinical Practice. New York: Raven Press, 1989:1-19.
49. van den Bosch H. Intracellular phospholipases A. Biochim Biophys Acta. 1980; 604: 191-246.
50. Awad AB, Ferger SL, Fink CS. Effect of dietary fat on the lipid composition and utilization of short-chain fatty acids by rat colonocytes. Lipids. 25:316-20, 1990.
51. Hillier K, Jewell R, Dorrell L, Smith CL. Incorporation of fatty acids from fish oil and olive oil into colonic mucosal lipids and effects upon eicosanoid synthesis in inflammatory bowel disease. Gut 32:1151-5, 1991.
52. Hecht JR and Heber D. Unpublished data.
53. Galli C, Marangoni F, Galella G. Modulation of lipid derived mediators by polyunsaturated fatty acids. Prostaglandins Leukotr Ess Fatty Acids. 1993;48:51-55.
54. Vane JR. Inhibition of prostaglandin synthesis as a mechanism of action for the aspirin-like drugs. Nature. 231:232-5, 1971.
55. Herschman HR, Fletcher BS, Kujubu DA. TIS10, a mitogen-inducible glucocorticoid-inhibited gene that encodes a second prostaglandin synthase/cyclooxygenase enzyme. J Lipid Mediat. 1993; 6:89-99.
56. Bennett A, Del Tacca M, Stamford IF, Zebro T. Prostaglandins from tumors of human large bowel. Br J Cancer. 1977; 35:881-4.
57. Jaffe BM, Parker AW, Philpott GW. Immunochemical measurement of prostaglandin or prostaglandin-like activity from normal and neoplastic cultured tissue. Surg Forum. 1971; 22:90-2.
58. Yamaguchi A, Ishida T, Nishimura G, Katoh M, Miyazaki I. Investigations of colonic prostaglandins in carcinogenesis in the rat colon. Dis Col Rectum. 1991; 34:572-6.
59. Earnest DL, Hixson LJ, Sim WW, Ahnen D, Alberts DS. Piroxicam as an agent to inhibit colonic neoplasia. Gastroenterology. 1989; 96:A134.
60. Neoptolemos JP, Husband D, Imray C, Rowley S, Lawson N. Arachidonic acid and docosahexanoic acid are increased in human colorectal cancer. Gut. 1991; 32:278-81.

61. Tsuji M and Dubois RN. Alterations in cellular adhesion and apoptosis in epithelial cells overexpressing prostaglandin endoperoxide synthase 2. Cell. 1995; 83:493-501.
62. Sano H, Kawahito Y, Wilder RL. Expression of cyclooxygenase-1 and -2 in human colorectal tumors. Cancer Res. 1995;55:3785-9.
63. Sato M, Narisawa T, Sano M, Takahashi T, Goto A. Growth inhibition of transplantable murine colon adenocarcinoma 38 by indomethacin. J Cancer Res Clin. Oncol 1983; 106: 21-6.
64. Olsson NO, Caignard A, Martin MS, Martin F. effect of indomethacin on the growth of colon cancer cells in syngeneic rats. Int J Immunopharmac. 1984; 6:329-334.
65. Hial V, De Mello MCF, Horakova, Beaven M. Antiproliferative activity of anti-inflammatory drugs in two mammalian cell culture lines. J Pharmacol Exp Ther. 1977; 202: 446-54.
66. Hubbard NE, Chapkin RS, Erickson KL. Inhibition of growth and linoleate-enhanced metastasis of a transplantable mouse mammary tumor by indomethacin. Cancer Lett. 1988; 43: 111-20.
67. Bayer BM, Kruth HS, Vaughan M, Beaven MA. Arrest of cultured cells in the G_1 Phase of the cell cycle by indomethacin. J Pharmol Exp Ther. 1979; 210: 106-11.
68. Pollard M, Luckert PH. Indomethacin treatment of rats with dimethylhydrazine-induced intestinal tumors. Cancer Treat Rep. 1980; 64:1323-7.
69. Narisawa T, Sato M, Tani M, et al. Inhibition of development of methylnitrosourea-induced rat colon tumors by indomethacin treatment. Cancer Res. 1981; 41:1954-7.
70. Pollard M, Luckert PH, Schmidt MA. The suppressive effect of piroxicam on autochthonous intestinal tumors in the rat. Cancer Lett. 1983; 21:57-67.
71. Reddy BS, Maruyama H, Kelloff G. Dose-related inhibition of colon carcinogenesis by dietary piroxicam, a nonsteroidal antiinflammatory drug, during different stages of rat colon tumor development. Cancer Res. 1987; 47:5340-6.
72. Moorghen M, Ince P, Finney KJ, et al. A protective effect of sulindac against chemically-induced primary colonic tumors in mice. J Pathol. 1988; 156: 341-7.
73. Skinner SA. Penney AG, O'Brien PE. Sulindac inhibits the rate of growth and appearance of colon tumors in the rat. Arch Surg. 1991; 126:1094-6.
74. Heneghan JB. Inhibition of human colon tumor growth in nude mice by indomethacin. In: Germfree Research: Microflora control and its application to the biomedical science. Alan Liss, 1985:315-8.
75. Rosenberg L, Palmer JR, Zauber AG. A hypothesis: nonsteroidal anti-inflammatory drugs reduce the risk of large-bowel cancer. J Natl Cancer Inst. 1991; 83: 355-8.
76. Thun MJ, Namboordi MM, Heath CW. Aspirin use and reduced risk of fatal colon cancer. N Engl J Med. 1991; 325:1593-6.
77. Suh O, Mettlin C, Petrelli NJ. Aspirin Use, Cancer, and Polyps of the Large Bowel. Cancer. 1993; 72:1171-7.
78. Giovannucci E, Rimm EB, Stampfer MJ, et al. Aspirin use and the risk for colorectal cancer and adenoma in male health professionals. Ann Intern Med. 1994; 121:241-6.
79. Kinzler KW, Nilbert MC, Su L-K, et al. Identification of FAP locus genes from chromosome 5q21. Science. 1991; 253:661-4.
80. Groden J, Thliveris A, Samowitz W, et al. Identification and characterization of the familial adenomatous polyposis coli gene. Cell. 1991; 66:589-600.
81. Waddell WR, Loughry RW. Sulindac for polyposis of the colon. J Surg Oncol. 1983; 24:83-7.
82. Waddell WR, Ganser GF, Cerise EJ, Loughry RW. Sulindac for polyposis of the colon. Am J Surg. 1989; 157: 175-9.
83. MacPhee M, Chepenik KP, Liddell RA, et al. The secretory phospholipase A2 gene is a candidate for the Mom1 locus, a major modifier of APC^{Min}-induced intestinal neoplasia. Cell. 1995;81:957-66.
84. Droy-Lefaix MT. Prostanoids and stomach physiology. and In: Curtis-Prior PB, ed. Prostaglandins: biology and chemistry of prostaglandins and related eicosanoids. Edinburgh; Churchill Livingstone. 1988; 345-60.
85. Morham SG, Langenbach R, Loftin CD, etal. Prostaglandin synthase 2 gene disruption causes severe renal pathology in the mouse. Cell. 1995;83:473-82.

13

HIGH FAT DIET, LIPID PEROXIDATION, AND PANCREATIC CARCINOGENESIS

Jianzhong Zhang[1] and Vay Liang W. Go[2]

[1] Research Fellow
Division of Clinical Nutrition
Department of Medicine
UCLA School of Medicine
[2] Professor of Medicine
Department of Medicine
UCLA School of Medicine
Los Angeles, California 90095

1. INTRODUCTION

Pancreatic cancer remains the fifth leading cause of death from cancer in the United States. It is the only cancer site where there is little difference between incidence and mortality rate (1). In terms of incidence, pancreatic cancer is now ranked ninth among cancer in the United States, with extremely bleak prognosis, and has a median survival time — regardless of therapy — of 2-3 months after diagnosis (1). Pancreatic cancer remains a significant clinical challenge in early diagnosis and effective management for cure of the disease. Research attention has now refocused on epidemiology and etiopathogenesis of pancreatic cancer, with the hope of developing effective measures for prevention and early diagnosis. Gordis and Gold (2) have reviewed the epidemiology and possible etiology of this disease. They concluded that both genetics and environment may play significant roles in the etiology of pancreatic cancer. Families at high risk for pancreatic cancer have been reported. Cigarette smoking is a risk factor, but the strength of the association is much less than in lung cancer. Animal studies have implicated several chemical exposures in pancreatic carcinogenesis. High dietary fat intake is implicated, while vitamins and/or dietary fiber may play protective roles against pancreatic cancer (3). The focus of this chapter is to review the evidence, based on animal studies, on the promotion of carcinogenesis by a high fat diet and on the protective effects of antioxidants in both chemically induced pancreatic cancer and through possible molecular mechanisms.

Dietary Fats, Lipids, Hormones, and Tumorigenesis, edited by Heber and Kritchevsky
Plenum Press, New York, 1996

2. DIETARY FATS

The influence of dietary fat on the post-initiation stage of carcinogenesis was first investigated by Tannenbaum in 1944 (4). In 1981, Robuck et al. (5) first investigated the role of dietary fats in pancreatic carcinogenesis in rats after injection of multiple doses of azaserine, a carcinogen used to produce pancreatic tumors in rats. High unsaturated fatty diets (20% corn oil, or 20% coconut oil) significantly increased the pancreatic neoplasm incidence during post initiation stage (5, 6). In these experiments, multiple doses of an initiating carcinogen were used, and the post initiation phase usually lasted 30-40 weeks. Since then, several studies confirmed the role of high fat diet, particularly diets containing considerable amounts of linoleic acid (7, 8, 9, 10), in the potentiation phase of carcinogenesis. Similar findings were reported in hamsters when a semi-synthetic diet, high in corn oil, significantly increased the incidence and number of pancreatic tumors induced by N-nitroso-bis (2-oxopropyl)amine (BOP). BOP produced tumors of ductal origin in hamsters which closely resemble the human cancer. After initiation with 20 mg BOP/kg body weight once weekly for 3 weeks, feeding hamsters with a high fat diet (20% lard), and feeding controls with 5% lard, significantly enhanced pancreatic carcinogenesis, in contrast to the post-initiation phase. Diets with high saturated and unsaturated fats have been shown to have no effect on the initiation phase of carcinogenesis induced by BOP in hamsters, or by azaserine in rats (11). Histologically, after 11-16 weeks of exposure, the hamster pancreas showed microscopic pre-neoplastic lesions (foci) of the pancreas, such as cystic, intermediate and tubular ductal complexes. A greater number of large ductal complexes in these three categories were found in hamsters fed on diets containing 20% saturated fat (lard), and were associated with an increase in the number of ducts with intraductal hyperplasia and adenocarcinoma. The promoting effects of both high lard diets and high corn oil diets on chemically induced pancreatic carcinogenesis were similar to those reported in experimental mammary (12) and colon carcinogenesis (13). Additionally reported is the minimum amount of essential fatty acids (4-8% of linoleic acid) in a 20% fat diet required to demonstrate pancreatic tumor promotion effects in rats (10) and hamsters (14). It has been postulated that the underlying mechanism might be related to increased intermediates of lipid peroxidation, which can increase the susceptibility of cellular DNA damage by certain carcinogens (15).

3. MECHANISM OF CARCINOGENESIS BY BOP

Foreign chemicals and carcinogens, like most drugs, require metabolic activation as well as detoxification by Phase I and II enzymes to their genotoxic/inactive intermediates. In some instances, these activated metabolites are subject to detoxification by conjugation via the various Phase II enzymes (16). Cytochrome P450s, hydroxylase, lipoxygenase and cyclooxygenases are Phase I enzymes. Phase II enzymes include UDP-glucuronosyl transferase and glutathione transferase, etc. Thus, the coordinate expression and regulation of Phase I and II drug-metabolizing enzymes and their metabolic balance in the cells of target organs may be an important factor in determining whether exposure to the carcinogen results in cancer development or not (16). Both BOP and azaserine require metabolism to an alkylating species that is mutagenic and capable of initiating a carcinogenic sequence yielding carcinomas of ductal and mixed acinar/ductal phenotypes, respectively. DNA methylation adducts were detected by the immunohistochemical method using mouse monoclonal antibody against O^6-methyl-deoxyguanosine and rabbit polyclonal antibody against 7-methyldeoxyguanosine (17). A correlation exists between the level of DNA alkylation and the relative carcinogenic potency of BOP and its active metabolites for their target tissues. Labeling of pancreatic ductal cells was greater and persisted longer than labeling of pancreatic acinar cells in an immunological study of tissue from BOP-induced hamsters(18). This observation could explain the selective targeting by BOP of ductal cells in the hamster pancreas, but similar labeling and persistence were also observed in kidney, where the incidence of tumor is low. It is possible the kidney, not pancreas, possesses certain detoxification enzymes that can eliminate the toxic effects of BOP locally.

Some drug-metabolizing enzymes are known to show genetic variability in their activity among individuals, which may explain their susceptibility to chemical carcinogenesis. The study of tobacco-related carcinogenesis, both *in vivo* and *in vitro*, indicates there is an association between the expression of certain Phase I enzymes, namely cytochrome P450 1A1, and the incidence of lung cancer among smokers (19). Cytochrome P450 1A1 is an enzyme known to activate certain xenobiotics into carcinogens, and it is observed that people with high cytochrome P450 1A1 enzyme activity are associated with high risks of lung cancer. This is probably due to a high metabolic activation of the carcinogens in cigarette smoke, such as polycyclic aromatic hydrocarbons, aromatic amine, nitrosamines, etc., into their active forms.

As in lung cancer, data on metabolic activation of carcinogens by cytochrome P450 isoforms in pancreas is increasing. It was found that cytochrome P450 1A1 /1A2 isoforms of hamster liver are predominantly involved in the metabolic activation of carcinogens, such as BOP, aflatoxin B1, and acetaminofluorene, etc. Interestingly, the hamster pancreas cannot activate BOP (20). BOP is also activated to mutagens by hepatic P450 isozymes from rats and mice (21). Immunohistochemical studies indicate that cytochrome P450 1A1 can be induced in fish pancreatic acinar cells and in ductal epithelium (22). This probably suggests that both acinar and ductal cells can activate BOP, although in this regard they both are less active when compared to hepatocytes. However, we did not find detectable enzyme activity of cytochrome P450 1A1 (measured by aryl hydrocarbon hydroxylase activity, unpublished data) in both immortalized pancreatic ductal cells TAKA-1 and in the pancreatic cancer cell line, PC-1, obtained from Dr. Pour et al. (23). Thus, our preliminary data and others' (20) support the suggestion that activated BOP metabolites are likely to transfer from liver to pancreas *in vivo* (24), and that the targeting of ductal cells for malignant transformation reflects biologic factors, for example, the transfer of activated species and slower DNA repair rather than the ability of the ductal cells to activate the carcinogen. It is indeed possible that an isoform of cytochrome P450 is involved in BOP activation and that ductal cells lose some of the enzyme activity during culturing (25). Nevertheless, why these chemicals specifically target pancreatic acinar and ductal cells needs further investigation.

4. LIPID PEROXIDATION AND METABOLISM OF BOP

There have been studies on the mechanism of polyunsaturated fatty acid (PUFA)-promoted effects on carcinogenesis, and the involvement of various reactive intermediates has been proposed. Oxidation of unsaturated fatty acids is the result of normal metabolism, physical activity, ionizing irradiation and metabolites of various chemicals, drugs and foods. The reactive oxygen species (ROS) generated during lipid peroxidation can be eliminated by the normal defense systems, such as enzymes and antioxidant nutrients. Under certain circumstances, if the ROS generated surpasses the capacity of the cellular defenses, oxidative damage will occur.

It is well known that ROS are generated during lipid peroxidation processes (26,27) or arise directly from linoleic acid hydroperoxide decomposition (28). Furthermore, alkoxy and peroxyl radicals, as well as reactive aldehydes, can be formed from PUFA hydroperoxides (29, 30, 31). Indeed, the induction of 7,8-dihydro-8-oxo-2′-deoxyguanosine, a marker of oxidative DNA damage, has been reported after exposure of cellular DNA to auto-oxidized methyl linoleate (32). Damage to DNA by oxygen-free radicals is frequently postulated to cause mutations that are associated with the initiation and progression of human cancers. However, it has been difficult to prove this relationship, due to the large number of ROS that have potential to damage DNA. Overviews of possible interactions of lipid peroxidation products with DNA are discussed elsewhere (33). Similarly, four DNA adducts (7-methylguanine, 7-hydroxypropylguanine, 0^6-methylguanine and 0^6-hydroxypropylguanine) were formed in pancreatic ductal cells following incubation with labeled N-nitroso(2-hydroxypropyl)(2-oxypropyl)amine (HPOP). HPOP is one of the two major reduced products of BOP *in vivo*; the other being N-nitrosobis(2-hydroxypropyl)amine (BHP) (25). Hepatocytes were 100 to 150 times more active than pancreatic duct cells in activating BOP (34). Higher concentrations of HPOP and BHP in pancreatic juice, rather than in circulation, were detected around 3 h after treatment, suggesting products of BOP are accumulated in the pancreas before dilution by circulation (25). Although there is no experimental evidence for the involvement of lipid peroxidation products in the promotion of carcinogenesis, the fact is that the most effective defense systems against mutagenic and carcinogenic effects also prevent oxygen radical-induced DNA damage and lipid peroxidation (35, 36). This suggests that the promotional effects of a high fat diet are due to their increased lipid peroxidation *in vivo*, "distracting" defense system efficiency against the carcinogenic effects of BOP.

5. FISH OIL, ARACHIDONIC ACID AND PROSTAGLANDIN

As discussed, in an experimental pancreatic cancer model, ω-6 PUFA (mainly linoleic acid) has been shown to promote carcinogenesis in the post-initiation phase (37, 38, 39, 40), and linoleic acid may give rise to prostaglandin (PGs) of the 2-series via arachidonic acid catalyzed by cycloxygenase. Some of these PGs stimulate cell proliferation *in vitro* ($PGF_2\alpha$), or have immunosupressive properties (PGE_2), and may promote tumor growth or prevent inhibition of tumor growth, respectively (41, 42). Inhibitors of PG-producing cyclooxygenase, such as indomethacin, have the ability to inhibit mammary carcinogenesis by 7,12-dimethyl-

benz[a]anthracene (DMBA) in rats fed high fat diets (43) and to decrease multiplicity of pancreatic tumors induced in hamsters by BOP (44). It is suggested that this anti-carcinogenic effect of cyclooxygenase inhibitors may be ascribed to inhibit PG formation. This mechanism may also play a role in the inhibitory effects of ω-3 PUFAs from fish oil on carcinogenesis (45). ω-3 PUFAs from fish oil, such as eicosapentaenoic acid (EPA), have been shown to inhibit tumor growth in experimental animal models (46). Competitive inhibition of formation of the 2-series PGs by ω-3 PUFAs may be responsible for the antitumor effects observed with dietary fish oil (45). It is now established that arachidonic acid release occurs in many kinds of oxidative stress, including heavy metals (47) and ozone (48). Arachidonic acid is converted to alcohols and epoxides by P450 1A1. Nebert (49) suggested that arachidonic acid release is involved with the oxidative stress response and that the release is associated with the expression of aryl hydrocarbon (Ah) receptor-inducible genes, which increases the expression of P450 1A1. It is very likely this increased P450 1A1 activity also increases prostaglandin synthesis. The mechanism of prostaglandin involvement needs further investigation.

6. OTHER ANTIOXIDANTS

In human epidemiological studies, other protective factors, such as vitamins C and E, B-carotene, glutathion, selenium, and high fiber diets, have been found to be associated with a lower risk of pancreatic cancer. However, these factors have not yet been systemically investigated in the animal model. These micronutrients contribute to cellular defense mechanisms against lipid peroxidation in microsomes, mitochondria, and nuclei (50). Particularly, the triads of α-tocopherol, ascorbate, and glutathione (reduced and oxidized forms) work together with enzymes, such as superoxide dismutase, catalase, and glutathione peroxidase, etc., and are able to "repair" oxidizing radicals (50, 51). An intervention study of such antioxidants is needed in a pancreatic tumor model in hamsters to see if the promotion effect of high fat diet can be prevented or delayed.

7. CONCLUSION

In conclusion, experimental studies have established that fatty acids can promote carcinogenesis in the pancreas. This confirms the observations in human epidemiological studies that high fat diets are risk factors in pancreatic cancer. The carcinogenic risk from exposure to environmental procarcinogens depends not only on the nature and dose of chemical exposure, but may also depend on the individual's defense system and sensitivity to the carcinogen (52, 53), summarised in Figure 1.

Reactive oxygen species occurs in tissues, and can damage DNA. Lipid peroxidation is a significant source of free radicals. The potentially deleterious reactions caused by increased lipid peroxidatioin are normally controlled in part by antioxidants that eliminate prooxidants and scavenge free radicals. However, during the postinitiation phase these reactions probably promote the carcinogenesis by reducing the defense mechanisms of the

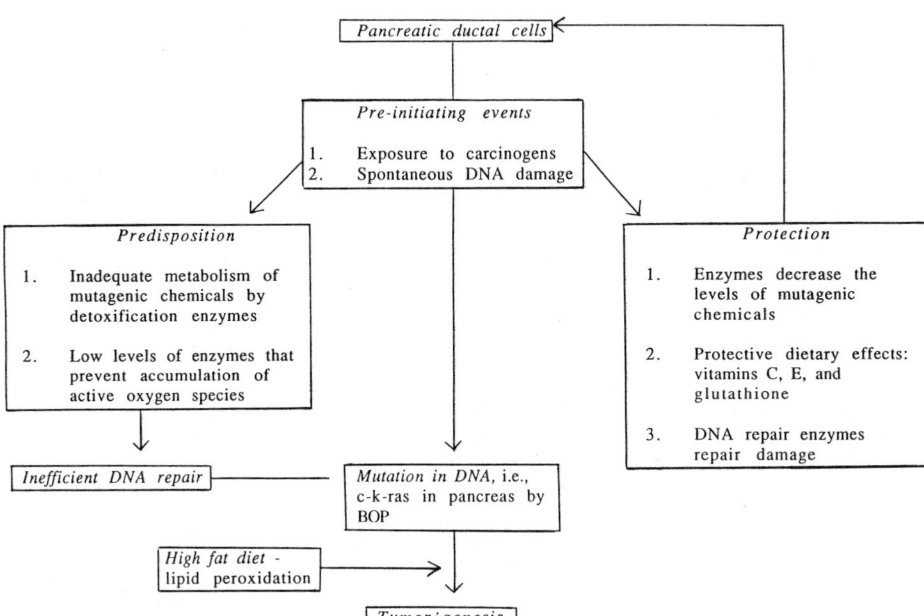

Figure 1. N-nitrosobis (2-oxopropyl)amine (BOP), dietary fat and other factors in pancreatic tumorigensis.

cells. Antioxidant functions are associatied with lowering DNA damage, malignant transformations, and other parameters of cell damage *in vitro*, as well as epidemiologically with lowered incidence of certain types of cancers, including pancreatic cancer.

Further experiments are needed to investigate the mechanisms of the promoting effects of fatty acids in tumorogenesis. Currently, various investigations are being conducted to determine if supplements of certain antioxidants, such as ascorbate, α-tocopherol, and glutathione, will delay or even protect the promotional effects of high fatty acids in chronically induced pancreatic carcinogenesis. Hopefully, these studies will provide useful information toward the beneficial uses of antioxidant nutrients in reducing risks for pancreatic cancer.

ACKWNOWLEDGMENT

Grateful ackwnowledgment is made to Anita Stein for her editorial assistance in the preparation of this chapter.

REFERENCES

1. Taylor WT, Everhart GE: Pancreatic cancer. In: *Digestive diseases in the United States: Epidemiology and impact*. Everhard EG, etc., USPHA, NIH Publication no. 49-492270, 1994.
2. Gordis L, Gold EB: Epidemiology and Etiology of Pancreatic Cancer, and in Go et al. (Ed.) *Pancreas, Biology, and Pathobiology of Disease* Second Edition, 1993, pp 837-855, Raven Press, N.Y.
3. National Cancer Institute: *Measures of progress against cancer*, 1982-1992. Bethesda: National Cancer Institute, 1994.
4. Tannenbaum A: The dependence of the genesis of induced skin tumors on the fat content of the diet during different stages of carcinogenesis. Cancer Res 1944; 4:683-687.
5. Roebuck BD, Yager, Jr. JD, Longnecker DS: Dietary Modulation of Azaserine-induced pancreatic Carcinogenesis in the Rat.Cancer Res 1981; 41:888-893.
6. Roebuck BD, Yager Jr. JD, Longnecker DS, Wilpone SA: Promotion by Unsaturated Fat of Azaserine-induced Pancreatic Carcinogenesis in the Rat. Cancer Res 1981; 41:3961-3966.
7. Longnecker DS, Roebuck BD, Kuhlmann ET: Enhancement of pancreatic carcinogenesis by a dietary unsaturated fat in rats treated with saline or N-nitroso (2-hydroxypropyl) (2-oxypropyl)amine. J Natl Cancer Inst 1985; 74:219-222.
8. O'Connor TP, Roebuck BD, Campbell TC: Dietary intervention during the postdosing phase of L-azaserine-induced preneoplastic lesions. J Natl Can Inst 1985; 75:955-957.
9. Roebuck BD, Longnecker, DS, Baumgartner, KJ, Thron, CD. Carcinogen-induced lesions in the rat pancreas: effects of varying levels of essential fatty acids. Cancer Res 1985; 45:5252-5256.
10. Roebuck BD, Kaplita PV, Edwards BR, Praissman M: Effects of dietary fats and soybean protein on azaserine-induced pancratic carcinogenesis and plasma cholecystokinin in the rat. Cancer Res 1987; 47:1333-1338.
11. Roebuck BD: Dietary Fat and the Development of pancreatic Cancer. Lipids 1992; 27:804-806.
12. Rogers AE, Lee SY: Chemically induced mammary gland tumors in rats: modulation by dietary fat. Dietary Fat and Cancer 1986, Alan R. Liss, Inc., New York, pp. 255-282.
13. Reddy BS: Diet and colon cancer: evidence from human and animal model studies. In Reddy BS, Cohen LA (eds). Diet, Nutrition, and Cancer: A Critical Evaluation. CRC Press 1986, Boca Raton; 1:47-65.
14. Appel, MJ, van Garderen-Hoetmer A, Woutersen RA: Effects of Dietary Linoleic Acid on Pancreatic Carcinogenesis in Rats and Hamsters. Cancer Res 1994; 54:2113-2120.
15. Reddy BS: Dietary Fat and Colon Cancer: Animal Model Studies. Lipids 1992; 27:807-812.
16. Nebert DW: Drug-Metabolizing Enzymes in Ligand-Modulated Transcription. Biochem Pharmacol 1994; 47:25-37.
17. Asamoto M, Mikheev AM, Jiang YZ, Wild CP, Hall J, Montesano R: Immunohistochemical detection of DNA alkylation adducts in rat and hamster liver after treatment with dimethylnitrosamine. Experimental Pathology 1991; 41:71-78.

18. Bax J, Pour PM, Nagel DL, Lawson TA, Woutersen RA, Scherer E; Long-term persistence of DNA alkylation in hamster tissues after N-nitrosobis(2-oxopropyl)amine. J. of Cancer Research and Clinical Oncology 1990; 116:149-155.
19. McLemore TL, Adelberg S, Liu MC, McMahon NA, Yu SJ, Hubbard WC, Czerwinski M, Wood TG, Storeng R, Lubet RA, et al: Expression of CYP1A1 gene in patients with lung cancer: evidence for cigarette smoke-induced gene expression in normal lung tissue and for altered gene regulation in primary pulmonary carcinomas. J. of the National Cancer Institute 1990; 82:1333-1339.
20. Mori Y, Iimura K, Fuukawa F, Nishikawa A, Takahashi M, Konishi Y: Effect of cigarette smoke on the mutagenic activation of various carcinogens in hamster. Mutation Res 1995; 346:1-8.
21. Mori Y, Yamazaki H, Toyoshi K, Denda A, Konishi Y: Mutagenic activation of carcinogenic N-nitro-so-propylamines by liver S9 fractions from mice, rats and hamsters: evidence for a cytochrome P450-dependent reaction. Carcinogenesis 1986; 7:375-379.
22. Husoy A-M, Myers MS, Willis ML, Collier TK, Celander M, Goksoyr A: Immunohistochemical Localization of CYP1A and CYP3A-like Isozymes in Hepatic and Extrahepatic Tissues of Atlantic Cod (Gadus morhua L.), a Marine Fish. Toxicol & Appl Pharmacol 1994; 129:294-308.
23. Egami, H, Takiyama Y,Cano M, Houser WH, Pour PM: Establishment of hamster pancreatic ductal carcinoma cell line (PC-1) producing blood group-related antigens. Carcinogenesis 1989; 10:861-869.
24. Longnecker DS: Carcinogenesis in the pancreas. Arch Pathol Lab Med 1983; 107:54-58.
25. Kokkinakis DM, Reddy MK, Norgle JR, Baskaran K: Metabolism and activation of pancreas specific nitrosamines by pancreatic ductal cells in culture. Carcinogenesis 1993; 14:1705-1709.
26. Cheeseman KH: Mechanisms and Effects of Lipid Peroxidation. Molec Aspects Med 1993; 14;191-197.
27. Comporti M: Lipid Peroxidation. Biopathological Significance. molec Aspects Med 1993. 14:199-207.
28. Yagi K, Ishida N, Komura S, Ohishi N, Kusai M, Kohno M: Generation of hydroxyl radical from linoleic acid hydroperoxide in the presence of epinephrine and iron. Biochem Biophys Res Commun (1992); 183:945-951.
29. Schreiber J, Mason PP, Eling TE: Carbon-centered free radical intermediates in the hematin- and vesicle-catalyzed decomposition of fatty acid hydroperoxides. Arch Biochem Biophys 1986; 251: 17-24.
30. Benamira M, Marnett L; The lipid peroxidation product 4-hydroxynonenal is a potent inducer of the SOS response. Mutat Res 1992; 293:1-10.
31. Vaca EC,Vodicka P, Hemminki K: Determination of malonaldehyde-modified 2′-deoxyguanosine-3′-monophosphate and DNA by ^{32}P-postlabelling. Carcinogenesis 1992; 13:593-599.
32. Park J, Floyd RA: Lipid peroxidaion products mediate the formation of 8-hydroxydeoxyguanosine in DNA. Free Radical Biol Med 1992; 12:245-250.
33. Vaca CE, Wilhelm J, Harms-Ringdahl M: Interaction of lipid peroxidation products with DNA. A review. Mutat Res 1987; 171:169-176.
34. Mangino MM, Scarpelli DG, Kokkinakis DM: Metabolism and activation of the pancreatic carcinogen N-nitrosobis(2-oxopropyl)amine by isolated hepatocytes and pancreatic cells of the Syrian hamster. Carcinogenesis 1990; 1:625-631.
35. Ames BN: Dietary carcinogens and anticarcinogens, oxygen radicals and degenerative diseases. Science 1983; 221:1256-1264.
36. Knuutila S: Role of free radicals in genetic damage (mutation). Med Biol 1984; 62:271-279.
37. Carroll KK: Experimental evidence of dietary factors and hormone dependent cancers. Cancer Res 1975; 35:3374-3383.
38. Cohen LA, Thompson DO, Maeura Y, Choi K, Blank ME, Rose DP: Dietary fat and mammary cancer I. Promoting effects of different dietary fats on N-nitrosomethylurea induced rat mammary tumorigenesis. J Natl Caner Inst 1986; 77:33-42.
39. Reddy BS, Maeura Y: Tumor promotion by dietary fat in azoxyethane-induced colon carcinogenesis in female F344 rats: influence of amount and sources of dietary fat. J Natl Cancer Inst 1984; 72:745-750.
40. Roebuck BD, Yager JD, Longnecker DS, Wilcore SA: Promotion by unsaturated fat of azaserine induced pancreatic carcinogenesis in the rat. Cancer Res 1981; 41:3961-3966.
41. Jimenez de Asua L, Otto AM, Lindgren JA, Hammarstrom S: The stimulation of the initiation of DMA synthesis and cell division in the Swiss mouse 3T3 cells by prostaglandin F_{2a} requires specific functional groups in the molecule. J Biol Chem 1983; 258:8774-8780.
42. Plescia OJ, Racis S: Prostaglandins as physiological immunoregulators. Prog Allergy 1988; 44:153-171.
44. Takahashi M, Furukawa F, Toyoda K, Sato H, Hasegawa R, Imaida K, Hayashi Y: Effects of various prostaglandin synthesis inhibitors on pancreatic carcinogenesis in hamsters after initiation with N-nitroso-bis(2-oxypropyl)amine. Carcinogenesis 1990; 11:393-395.
45. Appel MJ, Woutersen RA: Modulation of growth and cell turnover of preneoplastic lesions and of prostaglandin levels in rat pancreas by dietary fish oil. Carcinogenesis 1994; 15:2107-2112.

46. Karmali RA: Omega-3 fatty acids and cancer: a review. In Lands WEM (ed.), Proceedings of the AOCS Short Course on Polyunsaturated Fatty Acids and Eicosanoids. American Oil Chemists' Society, Champaign, IL 1988; 222-232.
47. Kafer A, Zoltzer H, Krug HF: The stimulation of arachidonic acid metabolism by organic lead and tin compounds in human HL-60 leukemia cells. Toxicology and Applied Pharmacology 1992;116:125-132.
48. McKinnon KP, Madden MC, Noah TL, Devlin RB: In vitro ozone exposure increases release of arachidonic acid products from a human bronchial epithelial cell line. Toxicology and Applied Pharmacology 1993; 118:215-223.
49. Nebert DW; Drug metabolism and signal transduction: Possible role of Ah receptor and arachidonic acid cascade in protection from ethanol toxicity. Exs 1994; 71:231-240.
50. Thomas CE, McLean LR, Parker RA, Ohlweiler DF: Ascorbate and Phenolic Antioxidant Interactions in Prevention of Liposomal Oxidation. Lipids 1992; 27:543-549.
51. Buettner GR: The Pecking Order of Free Radicals and Antioxidants: Lipid Peroxidation, a-Tocopherol, and Ascorbate. Arch Biochem & Biophys 1993; 300:535-543.
52. Omenn GS, Gelboin HV: Genetic variability in responses to chemical exposure. Banbury report 16. Cold Spring Harbor, NY: Cold Spring Harbor Laboratory, 1984:3-395.
53. Caporaso N, Landi MT, Vineis P: Relative of metabolic polymorphisms to human carcinogenesis: evaluation of epidemiological evidence. Pharmacogenetics 1991, 1:4-19.

14

NUTRITIONAL APPROACHES TO THE PREVENTION OF PROSTATE CANCER PROGRESSION

Ian Yip, William Aronson, and David Heber

Division of Clinical Nutrition
Department of Medicine and Division of Urology, Department of Surgery
UCLA School of Medicine
Los Angeles, California 90095

1. INTRODUCTION

Prostate cancer is becoming an increasingly important public health problem in the United States. As our population ages, the number of patients with clinically significant prostate cancer will be expected to increase. Prostate cancer is now the most commonly diagnosed cancer in men as well as the second leading cause of male cancer deaths in this country [1]. In 1995, an estimated 244,000 U.S. men will be diagnosed with prostate cancer and about 40,000 men will die of the disease [1]. Despite extensive efforts to study its pathogenesis, the origins of prostate cancer and the factors that promote its progression are not well established [2]. Nonetheless, there is a great deal of evidence suggesting that this disease is the result of genetic and environmental interactions. One of the most widely accepted risk factors for prostate cancer is the race-ethnicity background [3-4]. African-American men have the highest prostate cancer rates in the world, whereas Japanese and Chinese men native to their countries have the lowest rate of prostate cancer[4]. However, Chinese and Japanese men who immigrate to this country have been found to have an increased risk of developing clinical prostate cancers within a generation to approach that of the U.S. average [5]. In addition, the incidence of prostate cancer in Japan has been increasing at a time when Western diets and lifestyles are being adopted into that country[6-7]. These epidemiologic data suggest that the diet consumed in Western industrialized nations may be among one of the most important environmental factors which affect the development of prostate cancer. While latent or clinically insignificant prostate cancer occurs in large but equal rates in autopsy studies among men of Asian countries and the U.S. (approximately 30% of men older than age 50 years), the rate of clinically significant prostate cancer is 120-fold higher in the United States [1,8-9]. These data suggest that the same dietary factors may also promote the progression of latent or microscopic prostate cancer to clinically significant and metastatic prostate cancer [10]. While there is continuing debate on the impact of various treatment modalities; surgery, radiation, and hormonal therapies on the survival of patients with different stages

Dietary Fats, Lipids, Hormones, and Tumorigenesis, edited by Heber and Kritchevsky
Plenum Press, New York, 1996

of prostate cancer, the utilization of nutritional modification as a adjuvant therapy for treatment of prostate cancer is certainly an attractive idea. This chapter will explore evidences of the various nutritional approaches for the prevention of prostate cancer progression.

2. DIET, HORMONES AND LIPIDS IN PROSTATE CANCER

There is a substantial body of epidemiologic, experimental and metabolic studies providing evidence supporting the notion that diet may influence cancer development, progression, metastasis and mortality. Androgens are required for the normal growth, maintenance, and functional activity of the prostate gland, and prostatic cancer has been shown to be androgen-dependent. In animal studies, prostatic adenocarcinoma can be induced by prolonged administration of testosterone in rats [11]. For these reasons, the ablation of androgens from the circulation has formed the basis for the first-line therapy of metastatic prostate cancers [12]. Estradiol, an physiologically active estrogen, has been reported to be capable of inhibiting the growth of human prostate cancer cells in vitro [13]. It has been proposed that diet intervention may ultimately influence prostate cancer biology and its progression through its effects on the circulating levels of endogenous sex steroid hormones and lipids [3, 14].

2.1 Dietary Fat and Prostate Cancer

Epidemiologic studies have suggested a link between dietary fat intake and prostate cancer occurrence [15-20]. Essential fatty acids have been shown in vitro to affect cell proliferation, immune defenses, tissue invasiveness and metastatic spread of tumors as well as membrane fluidity which could affect cell surface receptors and cell-cell interactions [21]. Dietary fatty acids may be involved in the carcinogenic process within the prostate gland and progression to clinically manifested disease. In addition, changes in dietary fat content may affect prostate cancer growth through the effects of a specific type or group of fatty acids. It has also been shown that linoleic acid, an omega-6 polyunsaturated fatty acid exerted a stimulatory effect on the growth of an androgen-unresponsive human prostate cancer cell line, PC-3 in vitro[22]. Conversely, omega-3 fatty acids present in fish oils have been shown to inhibit PC-3 prostate cell growth in a dose-dependent manner. The activity of 5 alpha-reductase activity which converts testosterone into dihydrotestosterone within the prostate epithelial cell has also been shown to be reduced by several unsaturated fatty acids which include linolenic acid, palmitoleic acid and oleic acid [23]. In animal studies, it has been shown that feeding a high-fat diet before conception and from the beginning of organogenesis had a marked promotional effect on the early stage of prostate carcinogensis in ACI/seg rats [24]; and increasing amounts of dietary linoleic acid enhanced the growth of Dunning transplantable prostate adenocarcinomas in rats [11]. In addition, reduction of dietary fat has been shown to substantially slow the growth of tumors established from LNCaP human prostatic adenocarcinoma cells in athymic nude mice using a xenograft model [25].

Several epidemiologic studies have demonstrated a positive association between meat and milk consumption and prostate cancers [19-20, 26-27]. In humans, a retrospective study by West et al. [17] and a prospective study by Giovannucci et al. [18] suggested that it was the more aggressive prostate cancers that exhibited a significant correlation with high fat intake. The age-adjusted prospective study by Giovannucci et al. was highly significant for the highest quintile of meat intake with a risk of 2.64. The case-control study by West et al. analyzed the data separately for the age groups 45-67 and 68-74 years, and found a significant difference for tumor aggressiveness for the older age group in relation to fat intake less than

66 gram/d, 90-100 gram/d and over 120 g/d. Similarly, Kolonel [19] and co-workers, in a correlation study in Hawaii, found a significant relationship between dietary fat and prostate cancer mortality in men 70 years and older.

Although the link between dietary fat intake and prostate cancer has been one of the most consistently reproducible findings in the diet and cancer epidemiologic literature, the association has not yet been entirely convincing because most of these studies have their shortcomings. Not all investigators have selected an appropriate comparison groups, e.g. Western control groups do not consume a fat intake as low as in rural China or as in former years in Japan. Nutritional assessment in homogenous populations is difficult even when the macronutrients are examined, or such specific food items as meat and milk, which contribute approximately 40% and 20%, respectively, of the fat consumed by adults. Furthermore, there are problems of dietary assessment based on food frequency questionnaires which represent a limiting factor in sensitivity and specificity. In addition, some studies have had an insufficient sample size to accurately estimate the risk. While many epidemiologic studies have shown a positive association between cancer and dietary fat, the strength of the finding, in most, may have been diminished because patient age and tumor stage and aggressiveness were generally not taken into account. Yet, in spite of all of these limitations, one may conclude since several retrospective [19-20, 26-29] and prospective studies [16, 30-31] have found an association between prostate cancer and dietary fat, and none have shown a negative correlation, that the actual association may be quite high.

Hamalainen and colleagues [32] reduced the total fat consumption of healthy Finnish male volunteers aged 40-49 years from their customary level of 40% to 25% of total calories, and increased the ratio of dietary polyunsaturated to saturated fat. This intervention, which involved no change in energy intake and no deliberate effort to alter dietary fiber, produced significant decreases in both the total serum testosterone, and the non-protein-bound testosterone. In another group of normal males who were fed with a very low fat, low cholesterol and high-complex carbohydrate diet, their serum levels of sex steroid were also shown to be reduced [33]. Furthermore, in a group of Caucasian and African-American men fed a diet in which fat content was reduced from 40% to 30% of total calories, their urinary levels of androgens and estrogens were shown to decrease [34]. Given the stimulatory effects of androgens on established prostate cancer, it is plausible that an intensive dietary regimen which result in a decrease in specific androgens and estrogens or changes in ratios of sex hormones may impede the growth of prostate cancer partially or wholly on that basis.

An average American male consumes a diet where 35% of total calories are from fat. Evidence from ecologic data suggests that reduction in dietary fat calorie to 15% to 20% may be required in order to achieve a beneficial impact on the risk of fatal prostate cancer[35].

2.2 Soy Protein and Prostate Cancer

One of the major differences between Asian and American diets is the consumption of soy-based foodstuffs by Asians. In Taiwan, Coward et. al. reported the average consumption of soy protein is 35 g/day per capita [36]. The beneficial effects of soy have been attributed to the isoflavones found in highest content in soy, e.g. genistein and daidzein [37-38]. Genistein and Daidzein, and their beta-glucoside conjugates, are present in soybean in concentrations up to 3 mg/g, resulting in an intake of isoflavones of as high as 100 mg/day in this Taiwan population. The average isoflavone intake in the Asian culture is estimated to be 50 mg/day[37]. The average American only eats at most a few milligram of isoflavones per day.

The nine-fold decrease in prostate cancer mortality in Japanese men compared with those in the U.S. may be attributed, in part, to the high soy protein content in the Japanese diet [38]. Genistein and daidzein have been shown to inhibit the growth of androgen-dependent and androgen-independent prostatic cancer cell lines [39]. Akiyama et al. has shown that genistein is a potent inhibitor of epidermal growth factor receptor tyrosine kinase autophosphorylation in vitro [40]. In addition, genistein inhibits angiogenesis [41, 42] which is another critical requirement for tumor growth and metastasis [43]. Tumor angiogenesis, as measured by endothelial cell F8-RA staining and microvessel counting [44], correlates with invasive and metastatic prostate cancer [45]. The exact mechanism by which genistein exerts its inhibition on tumor angiogenesis is not well understood. It may act through attenuation of tyrosine protein kinases (TPK) [40]. It has been shown that genistein's inhibitory action on TPK results in a significant inhibition of the activity of cellular receptors for epidermal growth factor [46], insulin [47], insulin-like growth factor-1 [48], and platelet-derived growth factor [49]. In addition, genistein inhibits several other enzymes involved in signal transduction events that regulate cell growth, including inhibition of phosphatidylinositol turnover [50], topoisomerase II [51] and S6 kinase activity[52].

Enough scientific data have now accumulated to allow soy protein to be tested clinically in human trial as an adjuvant therapy for treatment of prostate cancer, however, the optimal and effective doses of soy protein to be used in humans will need to be further investigated.

2.3 Dietary Fiber and Prostate Cancer

Howie and Shultz [53] used 3-day food records to evaluate the dietary intakes of lacto-vegetarian and non-vegetarian Seventh-Day Adventists, and non-Seventh-Day Adventist omnivorous males aged 49-62 years. The three groups did not differ in their consumption of fat, but the vegetarians consumed more crude and dietary fiber; their plasma testosterone and estradiol levels were both significantly lower than those of the other two groups of men, and were inversely correlated with the estimates of dietary fiber intake. In this context, it is noteworthy that Rose and coworkers have also described an inverse relationship between the circulating estrogens and dietary fiber intake [54]. Decreased prostate cancer risk has also been found in Seventh Day Adventist men who have higher intakes of beans, lentils, and peas [55]. Since the amount of dietary fiber intake may influence circulating levels of testosterone and estradiol, dietary fiber may affect the prostate cancer biology by altering the circulating levels of sex steroids which in turn may decrease the progression of prostate cancer.

2.4 Selenium, Vitamin E and Prostate Cancer

Selenium is an essential trace element. An inverse relationship has been shown between selenium concentration in soil and mortality rate for bladder, lung, breast, and gastric cancer [56]. Epidemiological evidence also suggests that selenium may play a role in human cancer prevention [57-58]. Willet et al reported a strong association between low serum selenium level and prostatic cancer [59]. Selenium works synergistically with vitamin E to exert its effects [60]. Vitamin E is an anti-oxidant which acts by breaking the chain of hydrogen abstraction by peroxy radicals during the propagation phase of lipid peroxidation [61]. The combined effect of selenium and vitamin E has been shown to be more effective than each agent alone in inhibiting the growth of chemically-induced tumor in laboratory animals [62,63]. However, studies have not been performed to evaluate the effect of selenium alone or in combination with vitamin E against prostatic tumors in laboratory animal. Human studies designed to evaluate the potential chemopreventative effects of selenium and vitamin E are also lacking.

3. MEASURING THE EFFECT OF DIETARY INTERVENTION

Prostate cancer has a unique pattern of spread and tumor biology among hormone-dependent cancers. In order to demonstrate the effectiveness of nutritional interventions in retarding the progression of prostate cancer, it is necessary to carefully select the stage and degree of disease in which interventions will be advantagous. In addition, an intermediate marker is needed for monitoring effects of dietary intervention on progression of prostate cancer.

3.1 Prostate Specific Antigen as an Intermediate Marker

The sometimes long and variable natural history of prostate cancer has hindered structuring of prospective intervention studies. However, the identification of the prostate specific antigen (PSA) biomarker has greatly facilitated such trials [64]. PSA is a peptide produced by prostatic epithelial cells and prostate cancer cells [65, 66]. PSA has a role in early detection and staging of prostate cancer [67], but its most accurate and unequivocal application is in the follow-up of patients who have had radical prostatectomy for localized prostate cancer [68]. Following such surgery, if all of the cancer is erradicated, the PSA should become undetectable. Only in the extremely rare circumstance when benign prostatic tissue or prostate capsule has been inadvertently left in place by the surgeon does the PSA become detectable in the absence of tumor recurrence. In virtually every patient who has recurrence after surgery, the PSA becomes detectable and progressively rises before the tumor becomes clinically evident [69, 70]. Furthermore, recent data have demonstrated that more than 90 percent of patients who are destined to have clinical recurrence will be identified by a detectable PSA within three to five years of their radical prostatectomy [71-75]. PSA, therefore, is a reliable and sensitive indicator of tumor persistence and progression. With a detectable PSA as an indicator of tumor recurrence after radical prostatectomy, the exact probability of recurrence in various pathologic stages and grades has been documented in recent years.

3.2 Radical Prostatectomy for Prostate Cancer

The most common therapy in the United States for men with clinically localized prostate cancer and a life expectancy of at least 10 years is radical prostatectomy. Unfortunately, pathologic examination of the radical prostatectomy specimen reveals nonorgan- or nonspecimen-confined disease in 36% to 51% of patients [71-75]. Advanced pathologic findings in the radical prostatectomy specimen include tumor invasion through the prostate capsule, margin positive disease, and tumor extension into or metastatic to the seminal vesicles. In addition, patients with poorly differentiated tumors (combined Gleason score >7) are also more likely to have nonorgan- or nonspecimen-confined disease [75]. A number of institutions recently reviewed their results with radical prostatectomy for clinically localized prostate cancer and found that among the patients found to have nonorgan- or nonspecimen-confined disease, 25% to 50% of such patients have residual prostate cancer that was not removed by the radical prostatectomy, and with progression of this cancer go on to develop a detectable PSA within three years of their radical prostatectomy, and develop clinical prostate cancer (local or metastatic disease) 1 to 4 years after the PSA becomes detectable [71-75]. Patients who are at a high risk for having residual disease after radical prostatectomy, the majority of whom will have an undetectable PSA immediately postoperatively, are an ideal patient population to study experimental adjuvant therapies, with disease recurrence and progression measurable by a detectable PSA.

4. CONCLUSION

An important public health initiative now exists to identify nontoxic therapies to treat malignancy. Randomized intervention trials of manupulated dietary differences are needed to identify not only specific causal associations between diet and disease but also the safety of the tested dietary modifications and whether they can lower community-wide rates of disease. Investigators are now designing clincial trials to determine whether intensive nutritional modifications which utilize various interventions such as low fat, high fiber, soy protein, selenium, vitamin E alone or in combination can delay progression of residual prostate cancer in patients who have had a radical prostatectomy for adenocarcinoma of the prostate and are at a high risk for recurrence. In the future, if intensive nutrition intervention is found to delay prostate cancer progression, dietary modification could potentially be used as an adjuvant nontoxic therapy for patients who have had curative treatments for clinically localized prostate cancer and are at a high risk for having residual disease and subsequent biochemical and clinical failure. If adjuvant nutritional therapy can decrease the progression of residual micrometastatic disease for a sufficiently long period in patients who have had radical prostatectomy, this may spare patients from receiving adjuvant radiation therapy or androgen deprivation therapy and the morbidity associated with these treatments, and ultimately improve the quality of their life. In addition, based on the results of several studies on expectant management for men with clinically localized prostate cancer, many patients are electing not to undergo aggressive therapy for their prostate cancer and are opting for watchful waiting. These men could potentially benefit from nutritional therapy to slow the progression of their prostate cancer. Finally, if adjuvant nutritional therapy is found to delay the progression of residual prostate cancer, this may then have implications in regards to testing nutritional chemoprevention of prostate cancer in the general population.

REFERENCES

1. Wingo PA, Tong T, Bolden S. Cancer statistics. CA Cancer J Clin 1995; 45:8-30.
2. Carter HB, Coffey DS. The prostate: an increasing medical problem. Prostate 1990; 16:39-48.
3. Ross RK, Henderson BE. Do diet and androgens alter prostate cancer risk via a common etiologic pathway? J Natl Cancer Inst 1994; 86:252- 254.
4. Waterhouse J, Muir C, Shanmugaratnam K, et al. Cancer incidence in five continents. In Cancer Incidence, vol. 6 Lyon, France IARC Publ. No. 42, 1982.
5. Mandel JS, Schuman LM. Epidemiology of cancer of the prostate. Rev Cancer Epidemiol 1980; 1:1-65
6. Boyle P, Kevi R, Lucchuni F, La Vecchia C. Trends in diet-related cancers in Japan: a conundrum? Lancet 1993; 342:752.
7. Wynder EL, Fujita Y, Harris RE, Hirayama T, Hiyama T. Comparative epidemiology of cancer between the United States and Japan. A second look. Cancer 1991; 67:746-763.
8. Yatani R, Chigusa I, Akazaki K, Stemmermann GN, Welsh RA, Correa P. Geographic pathology of latent prostatic carcinoma. Int J Cancer 1982; 9:611-616
9. Parkin DM, Muir CS, Whelan SL, et al. Cancer incidence in five continents. Vol 6. IARC Sci Publ No. 120. Lyon:IARC 1992:970-971.
10. Wynder EL, Rose DP, Cohen LA. Nutrition and prostate cancer: a proposal for dietary intervention. Nutr and Cancer An International Journal 1994; 22:1-10.
11. Clinton SK, Palmer SS, Spriggs CE, Visek WJ. Growth of the Dunning transplantable prostate adenocarcinoma in rats fed diets with various fat contents. J Nutr 1988; 118:908-914.
12. Nesbit RM, Baum WC. Endocrine control of prostatic carcinoma: clinical and statistical survey of 1818 cases. JAMA 1950; 143:1317.
13. Carruba G, Pfeffer U, Fecarotta E et al. Estradiol inhibits growth of hormone-nonresponsive PC3 human prostate cancer cells. Cancer Res 1994; 54:1190-1193.
14. Pienta KJ, Esper PS. Risk factors for prostate cancer. Annals of Internal Medicine 1993; 118:793-803.

15. Gann PH, Hennekens CH, Sacks FM, Grodstein F, Giovannucci EL. Prospective study of plasma fatty acids and risk of prostate cancer. J Natl Cancer Inst 1994; 86:281-286.
16. Rose DP, Connolly JM. Dietary fat, fatty acids and prostate cancer. Lipids 1992; 27:798-803.
17. West DW, Slattery ML, Robison LM, French TK, and Mahoney AW. Adult dietary intake and prostate cancer risk in Utah: a case-control study with special emphasis on aggressive tumors. Cancer Causes Control 1991; 2:85-94.
18. Giovannucci E, Rimm EB, Colditz GA, Stampfer MJ, Ascherio A, Chute CC, Willett WC. A prospective study of dietary fat and risk of prostate cancer. J Natl Cancer Inst 1993; 85:1571-1579.
19. Kolonel LN, Yoshizawa CN, and Hankin JN. Diet and prostatic cancer: a case-control study in Hawaii. Am J Epidemiol 1988; 127:999-1012.
20. Mettlin C, Selenskas S, Natarajan N, and Huben R. Beta-carotene and animal fats and their relationship to prostate cancer risk. Cancer 1989; 64:605-612.
21. Karmali RA. Eicosanoids in neoplasia. Prev Med 1987; 16:493-502.
22. Rose DP, Connolly JM. Effects of fatty acids and eicosanoid synthesis inhibitors on the growth of two human prostate cancer cell lines. Prostate 1991; 18:243-254.
23. Liang T, Liao S. Inhibition of steroid 5 alpha-reductase by specific aliphatic unsaturated fatty acids. Biochem J. 1992; 285:557-562.
24. Kondo Y, Homma Y, Aso Y, Kakizoe T. Promotional effect of two- generation exposure to a high-fat diet on prostate carcinogenesis in ACI/Seg rats. Cancer Res 1994; 54:6129-6132.
25. Wang Y, Corr JG, Thaler HT, Tao Y, Fair WR, Heston WD. Decreased growth of established human prostate LNCaP tumors in nude mice fed a low-fat diet. J Natl Cancer Inst 1995; 87:1456-1462.
26. Kaul L, Heshmat MY, Kovi J et al. The role of diet in prostate cancer. Nutr Cancer 1987; 9:123-128.
27. Snowdon DA, Phillips RL, and Choi W. Diet, obesity, and risk of fatal prostate cancer. Am J Epidemiol 1984; 120:244-250.
28. Ross RK, Shimizu H, Paganini-Hill A, Honda G, Henderson BE. Case- control studies of prostate cancer in blacks and whites in southern California. J Natl Cancer Inst 1987;78:869-874.
29. Talamini R, LaVecchia C, Decarli A, Negri E and Franceschi S. Nutrition, social factors and prostatic cancer in a Northern Italian population. Br J Cancer 1986; 53:817-821.
30. Severson RK, Nomura AM, Grove JS, Stemmermann GN. A prospective study of demographics, diet, and prostate cancer among men of Japanese ancestry in Hawaii. Cancer Res 1989; 49:1857-1860.
31. Hsing AW, McLaughlin JK, Schuman LM, Bjelke E, Gridley G, Wacholder S, Chien HT, Blot WJ. Diet, tobacco use, and fatal prostate cancer: results from the Lutheran Brotherhood cohort study. Cancer Res 1990; 50:6836-6840.
32. Hamalainen E, Adlercreutz H, Pusak P, and Pietinen P. Diet and serum sex hormones in healthy men. J Steroid Biochem 1984; 20:459-464.
33. Rosenthal MB, Barnard RJ, Rose DP, Inkeles S, Hall J, Pritikin N. Effects of a high-complex-carbohydrate, low fat, low cholesterol diet on levels of serum lipids and estradiol. Am J Med 1985; 78:23-27.
34. Hill P, Wynder EL, Garbaczewski L, Walker AR. Diet and urinary steroids in black and white North American men and black South African men. Cancer Res 1987; 47:2982-2985.
35. Wynder EL, Weisburger JH, Ng SK. Nutrition: the need to define "optimal" intake as a basis for public policy decisions. Am J Public Health 1992; 82:346-350.
36. Coward L, Barnes NC, Setchell KDR, Barnes S. Genistein, diadzein and their β-glycoside conjugates: antitumor isoflavones in soybean foods from American and Asian diets. J Agric Food Chem 1993; 41:1961-1967.
37. Messina MJ, Persky V, Setchell KD, Barnes S. Soy intake and cancer risk: a review of the in vitro and *in vivo* data. Nutrition and Cancer 1994; 1:113-131.
38. Messina M, Barnes S. The role of soy products in reducing risk of cancer. JNCI 1991; 83:541-546.
39. Peterson G, Barnes S. Genistein and biochanin A inhibit the growth of human prostate cancer cells but not epidermal growth factor receptor authophosphorylation. Prostate 1993; 22:335-345.
40. Akiyama T, Ishida J, Nakagawa S, Ogawara H, Watanabe S, Itoh N, Shibuya M, Fukami Y. Genistein, a specific inhibitor of tyrosine specific protein kinases. J Biol Chem 1987; 262:5592-5595.
41. Schweigerer L, Christeleit K, Fleischmann G, Adlercreutz H, Wahala K, Hase T, Schwab M, Ludwig R, Fotsis T. Identification In human urine of a natural growth inhibitor for cells derived from solid paediatric tumours. Euro J Clin Invest 1992; 2:260-264.
42. Fotsis T, Pepper M, Adlercruetz H, Fleischmann G, Hase T, Montesano R, Schweigerer L. Genistein, a dietary-derived inhibitor of in vitro angiogenesis. Proc Natl Acad Sci USA 1993;90:2690-2694.
43. Folkman J. What is the evidence that tumors are angiogenesis dependent? J Natl Cancer Inst 1990; 82:4-6.
44. Brem H, Folkman J. Analysis of experimental antiangiogenic therapy. J Pediatr Surg 1993; 445-451.

45. Weidner N, Carroll PR, Flax J, Blumenfeld W, Folkman J. Tumor angiogenesis correlates with metastasis in invasive prostate carcinoma. Am J Pathol 1993; 143:401-409.
46. Linassier C, Pierre M, Le Pecq JB, Pierre J. Mechanisms of action in NIH- 3T3 cells of genistein, an inhibitor of EGF receptor tyrosine kinase activity. Biochem Pharmacol 1990; 39:187-193.
47. Kasuga M, Zick Y, Blithe DL, Crettaz M, Kahn CR. Insulin stimulates tyrosine phosphorylation of the insulin receptor in a cell-free system. Nature 1982; 298:667-669.
48. Rubin JB, Shia MA, Pilch PF. Stimulation of tyrosine-specific phosphorylation in vitro by insulin-like growth factor I. Nature 1983; 305:438-440.
49. Nishimura J, Huang JS, Deuel TF. Platelet-derived growth factor stimulates tyrosine-specific protein kinase activity in Swiss mouse 3T3 cell membranes. Proc Natl Acad Sci USA 1982;79:4303-4307.
50. Imoto M, Yamashita T, Sawa H, Kurasawa S, Naganawa H, Takeuchi T. Inhibition of cellular phosphatidylinositol turnover by psitectorigenin. Febs Lett 1988; 230:43-46.
51. Okura A, Arakawa H, Oka H, Yoshinari T, Monden Y. Effect of genistein on topoisomerase activity and growth of [val 12] Ha-ras- transformed NIH 3T3 cells. Biochem Biophys Res Commun 1988; 157:183-189.
52. Wei H, Wei L, Frenkel K, Bowen R, Barnes S. Inhibition of tumor promoter-induced hydrogen peroxide formation in vitro and *in vivo* by genistein. Nutr. and Cancer 1993; 20:1-12.
53. Howie BJ, and Shultz TD. Dietary and hormonal interrelationships among vegetarian and Seventh-Day adventists and nonvegetarian men. Am J Clin Nutr 1985:42:127-134.
54. Rose DP, Goldman M, Connolly JM, and Strong LE. A high-fiber- supplemented diet reduces serum estrogen levels in premenopausal women. Am J Clin Nutr 1991; 54:520-525.
55. Mills PK, Beeson WL, Phillips RL, Fraser GE. Cohort study of diet, lifestyle, and prostate cancer in Adventist men. Cancer 1989; 64:598-604.
56. Schrauzer GN, White DA, Schneider CJ. Cancer mortality correlation studies III. Statistical associations with dietary selenium intakes. Bioinorg Chem 1977; 7:23-34.
57. Hunter DJ, Morris JS, Stampfer MJ, Colditz GA, Speizer FE, Willett WC. A prospective study of selenium status and breast cancer risk. JAMA 1990; 264:1128-1131.
58. Van den Brandt PA, Goldbohm RA, van't Veer P, Bode P, Dorant E, Hermus RJJ, Sturmans F. A prospective cohort study on selenium levels and risk of gastrointestinal cancer. J Natl Cancer Inst 1993; 85:224-29.
60. Fox JM. Selenium: Nutritional implications and prospects for therapeutic medicine. Methods and Findings in Exp and Clin Pharm 1992; 14:275-287.
61. Cotgreave JA, Moldus P, Orrenius S. Host biochemical defense mechanisms against peroxidants. Ann Rev Pharm Toxicol 1988; 28:189-212.
62. Horvath P, Ip C. Synergistic effect of vitamin E and selenium in the chemoprevention of mammary carcinogenesis in rats. Cancer Res 1983; 43:5335-5341.
63. Takeda H, Hiroka T, Hatano T, Hamada Y, Yamamoto M. Inhibition of 7,12-dimethylbenz[a]anthracene-induced lipid peroxidation and mammary tumor development in rats by vitamin E in conjunction with selenium. Nutr Cancer 1992; 17:115-122.
64. Wang MC, Valenquela LA, Murphy GP, Chu TM. Purification of a human prostate specific antigen. Invest Urol 1979;17:159-163.
65. Papsidero LD, Kuriyama M, Wang MC, Horoszewicz J, Leong SS, Valenquela L, Murphy GP, Chu TM. Prostate Antigen: a marker for human prostate epithelial cells. J Natl Cancer Inst 1981; 66:37-42.
66. Nadji M, Tabei SZ, Castro A, Chu TM, Murphy GP, Wang MC, Morales AR. Prostatic-specific antigen: an immunohistologic marker for prostatic neoplasms. Cancer 1981; 48:1229-1232.
67. Catalona WJ, Smith DS, Ratliff TL, Dodds KM, Coplen DE, Yuan JJ, Petros JA, Andriole GL. Measurement of prostate-specific antigen in serum as a screening test for prostate cancer. NEJM 1991; 324:1156-1161.
68. Stamey TA, Yang N, Hay AR, McNeal JE, Freiha FS, Redwine E. Prostate-specific antigen as a serum marker for adenocarcinoma of the prostate. NEJM 1987; 317:909-916.
69. Stein A, deKernion JB, Smith RB, Dorey F, Patel H. Prostate specific antigen levels after radical prostatectomy in patients with organ confined and locally extensive prostate cancer. J Urol 1992; 147:942-946.
70. Brawer MK. Prostate-specific antigen: critical issues. Urology 1994; 44(6A):9-17.
71. Trapasso, JG, deKernion JB, Smith RB, Dorey F. The incidence and significance of detectable levels of serum prostate specific antigen after radical prostatectomy. J Urol 1994; 152:1821-1825.
72. Paulson DF. Impact of radical prostatectomy in the management of clinically localized disease. J Urol 1994; 152:1826-1830.
73. Walsh PC, Partin AW, Epstein JI. Cancer control and quality of life following anatomical radical retropubic prostatectomy: Results at 10 years. J Urol 1994; 152:1831-1836.

74. Catalona WJ, Smith DS. 5-year tumor recurrence rates after anatomical radical retropubic prostatectomy for prostate cancer. J Urol 1994;152:1837-1842.
75. Ohori M, Goad JR, Wheeler T,. Eastham JA, Thompson TC, Scardino PT. Can radical prostatectomy alter the progression of poorly differentiated prostate cancer? J Urol 1994; 152:1843-1849.

INDEX